LANDSCAPE ECOLOGY

A Top-Down Approach

Edited by

Jim Sanderson, Ph.D.
Larry D. Harris, Ph.D.

LEWIS PUBLISHERS
Boca Raton London New York Washington, D.C.

Library of Congress Cataloging-in-Publication Data

Catalog record is available from the Library of Congress.

Visit the CRC Press Web site at www.crcpress.com

© 2000 by CRC Press LLC
Lewis Publishers is an imprint of CRC Press LLC

No claim to original U.S. Government works
International Standard Book Number 1-56670-368-9
Library of Congress Card Number 99-40288
Printed in the United States of America   3 4 5 6 7 8 9 0
Printed on acid-free paper

Preface

The discipline of Landscape Ecology is rapidly emerging as a motive force, both in the domain of theoretical ecology, and in applied fields such as biodiversity conservation planning. Without it and its further development, the more reductionist elements of, and approaches to, ecology will continue to make the discipline decreasingly relevant to land management, which will be made totally on the basis of politics and socioeconomics.

Already, outstanding landscape ecology authorities of northern Europe assert that humans are a major component of all landscape ecology. Yes, in Saskatchewan, Sweden, and southern Siberia where modern humans have existed for about as long as there has been land not covered by glacier, this might be true. But anyone who has lived on and/or studied landscape ecology of the Sahara, Kalahari, or Patagonia would probably not view humans as being quite so important, relative to other biotic and abiotic forces of nature. We do not want to leave the impression that either classical ecology or modern humans are unimportant; rather, we assert that the relative degree of importance of raw, physical, pre-humanoid ecology compared to the role of politicians, engineers, or agriculturalists varies a great deal over different parts of earth.

It is also in this vein that we acknowledge that humans are probably the only creatures on earth that appreciate landscape ecology functions, however they represent a far cry from the majority of creatures who depend upon and/or control it. In other words, landscape seems to be a wonderful concept with which humans concern themselves. Landscape architecture, landscape beauty, and the humbling nature of landscape massivity are all wonderful human issues, but this work deals only with the ecology of landscapes, not its beauty or grandeur as perceived by humans.

Nevertheless, humans enter the picture in another way, perhaps perversely. Only humans study ecology and other disciplines and thus any and all theory, concept, or principle is a figment of the human brain. Humans have developed ecological theory and construct based upon the push and shove of competing ideas and ways of looking at the universe around them. This provides, we believe, one of the strongest cases for drawing distinction between approaches and bodies of knowledge such as that associated with ecosystem ecology vs. landscape ecology.

One aspect of landscape ecology that we believe to be most importantly distinct from other fields of ecology is that it explicitly encompasses and builds upon the role of heterogeneity in space as well as time. This is contrary to most ecological theory, concept, and principle that has been constructed over the last century. The reductionist, Cartesian approach to the creation of knowledge has become increasingly important in ecology over the last 40 years, and by taking this approach, it seems almost inevitable that underlying assumptions such as within-group homogeneity (be it population, community type, or biome), the assertion that hierarchy is critical to understanding various levels of organization, and the importance of defining, measuring, and building theory around the notion of discrete entities come to dominate the paradigms. Systems science handles all of these notions very easily by allowing the investigator to arbitrarily define the boundaries of the system under question and proceed to treat the various forces as either endogenous or exogenous to the system.

Because landscapes cover large areas and are, at least by our definition, heterogeneous, they fall outside the classical and preferred domain of scientific disciplines predicated upon the slicing and dicing of the Cartesian approach to knowledge. Notions such as hierarchy and discrete packaging (taxonomy) that are so deeply entrenched in Western thought may need to compromise as we explicitly seek to study the role of heterogeneity rather than discrete, supposedly homogeneous units that have dominated ecological thought for the last 100 years.

Many powerful forces of great significance to landscape ecology may not be served well by forcing them into a hierarchical mind set. Landscape ecology may or may not be aided by continued obligatory dependence upon ideas of the diversity of life being classifiable into neat packages referred to as species, genera, or what have you. For example, in spite of the fact that gradient analysis has been available as a paradigm for diversity, ecology still relies upon the notion that very diverse things such as the color spectrum must be pushed into namable packages before diversity can be calculated. And so, we ask the question, "How will future landscape ecologists calculate the diversity of a rainbow, or a gradient of vegetation across a large space?"

For at least 75 years, applied scientists such as game and fisheries managers have accepted that the most fruitful spatial areas for investigations occurred at the edges or interfaces of biological systems. Fisheries productivity in estuaries, hunting along field borders, and deep thought about why so much biotic activity (and so many marine mammals) occurred at the edges and vortices of the Gulf Stream, constitute the fundamentals of some disciplines. However, rigidity of paradigm prevailed and the discipline of ecology continued to focus on the study of supposedly more homogeneous-than-not, discrete units that were organized in a hierarchical structure such as populations, species, and communities. In the same vein, American ecologists have, until recently, insisted upon the study of identifiable types of (1) "natural areas" that were (2) devoid of humans, and (3) could be studied and/or saved as discrete patches. Only after a full century of reliance upon these fundamental thoughts do we approach the new millennium with quite a different mindset.

In our view, landscape ecology is different in that it not only explicitly recognizes heterogeneity, but also embraces and puts major emphasis upon the spatially explicit nature of phenomena. It not only recognizes that humans exist, it deals with them explicitly as entities and forcing functions on the landscape. The concept of place is important, to be sure, but the concept of space is arguably more important. Moreover, the concept of space being empty is increasingly at odds with the facts. Listen in on the next strategy session of a group of transportation engineers discussing the strategic plan for Wyoming or western Australia and the grating phrase will persistently irritate. And so, until which time as the major ecological entities and/or processes are shown to be homogeneous, the inevitability of landscape ecology must be accepted.

The pattern by which this new discipline unfolds and solidifies its pyramid of concept and principle remains to be seen. It goes without saying that many important ecological processes of mid-latitude ponds and lakes are fundamentally different from those of the surrounding terrestrial systems. Similarly, both the structural and functional variables important to the characterization of a forest are fundamentally different from those of a surrounding prairie or grassland ecosystem. Thus, to understand and truthfully articulate the ecology of an oak-hickory forest, we must not only find and define it in the ideal, we must map, describe, and quantify the cause-effect relations going on there. Furthermore, not only must we study the forest, or the patch of it, we must study the cross-boundary fluxes of energy, information, and materials, what systems analysts refer to as inputs and outputs. Harris (1984, and elsewhere) describes this as the issue of Content vs. Context, a theme that has become excruciatingly compelling as humans continue to slice and dice the continuity of natural systems as though we were all students in a global class studying comparative anatomy. This approach is overwhelming for some who are old enough to have measured the ecological and human-service roles that expansiveness and connectivity played in former landscapes. To some, it is downright cruel to be forced to bear witness to the insidious erosion of initial ecological integrity and biodiversity of one state park or another that is neither small enough to outfit with tennis courts and merry-go-rounds, nor large enough to contain any of the larger denizens that performed such crucial ecological and human-service functions. Others seem to enjoy watching the grass grow.

To be sure, some forests are larger than others, and without question the ecological relations of a small patch of forest may be quite different from those of a much larger forest of the same type and species composition that is surrounded by an orchard, a grove, a tree plantation, or a woodland. Moreover, this begs the question of scale. Some basic ecosystem and landscape properties are clearly scale-independent. In other words, there is no compelling reason why an oak tree growing alone in a yard or with two others in a forest patch should be any different from an identical twin growing within the sanctity of a large forest. On the other hand, it is equally obvious that other, critically important, structural and functional attributes are extremely scale-dependent. No serious thinker believes that a U.S. dollar, when possessed alone, is worth the same amount as a similar dollar accompanied by a million others. The same principle often applies to population viability, edge-to-area and volume-to-surface ratios, as well as the size of the ecosystems one chooses to study. The effect of forest productivity and/or respiration on the ambient carbon dioxide levels within the stand most assuredly depend upon the size and degree of ventilation of the patch. The effect and/or desirability of phenomena such as lightning strikes and hurricanes, life or death matters for many humans on earth, are directly scale-dependent upon the size of the research plot being studied.

To be sure, some boundaries between systems are much more clearly and easily definable than others. Any sentient creature occupying a spot on top of a precipice of high-latitude rocky coast will likely sense that a bit of movement in one direction would involve the ocean while a bit of movement in the opposite direction would involve the land. Some boundaries between ecosystems can be very sharp. And this not only applies to boundaries in two-dimensional space, but slight changes in time or other contextual variables as well. On the other hand, all boundaries are no less important just because they are more fuzzy. The admixture of fresh and salt water referred to as an estuary is no less real or important just because it is broader. Indeed, personnel of Everglades National Park would be the first to admit that the boundary between terrestrial and marine is pretty fuzzy under the sharpest of conditions. Two decades of legal bickering over wetlands boundaries testify to the complexity of what is involved here. By the same token, as the sea level continues to rise what does it matter (in law) that your grandfather owned a wonderful house with a great view on one of the barrier islands that formerly existed off the coast of Louisiana or panhandle of Florida? In one of his scores of seminal works Odum (1971) observed:

> Any unit that includes all of the organisms (i.e., the "community") in a given area interacting with the physical environment so that a flow of energy leads to clearly defined trophic structure, biotic diversity, and material cycles (i.e., exchange of materials between living and nonliving parts) with the system is an ecological system or ECOSYSTEM.

We see no reason to quibble.

However, for the very simple reason that the term and the concept of ecosystem has been so lucidly (and effectively) defined, why then would we overtly sully the term's meaning and utility by asserting that a landscape is simply a larger-version ecosystem. We do not. For pure and simple logic and lucidity we define landscape as consisting of two or more ecosystems in close proximity. Harris et al. (1996) asserted that an ecosystem can be any size ranging from a rotting log to the biosphere itself, and that a landscape is simply a "largish" ecosystem. Although we accept that word definitions can and do change with time, we also teach that for the purpose of communication, the power of a word is proportional to its specificity definitiveness. It seems tautological that any word, or concept, that can mean anything, does of necessity, mean nothing! Thus, for both heuristic and practical reasons we assert that the word and concept of "landscape" must be explicitly defined as distinct from ecosystem, not left dangling in the minds of devious agents of society, or agency administrators, to pick and choose and fight billion dollar lawsuits over nuance.

One of our highly esteemed colleagues and gentleman friends was asked in a job interview what he thought about the emerging field of landscape ecology. Aside from the fact that he may have wanted to please, his answer ("Well, it is simply a matter of scale isn't it?") was quick, assertive, and has stood as a dictum ever since, at least in his lab. We do not believe this is true. Moreover, the convenient classroom mantra, (which we have previously used) "it all depends upon the scale" is equally flawed if one studies closely and intensely. As stated above, it seems that some variables are highly scale-dependent, while others are not.

The discipline of ecology has advanced at an incredible rate, albeit on a slightly different course from that originally conceived (Clark 1973). This has been accompanied by a proliferation of highly useful subdisciplines ranging from physiological ecology to global-systems ecology. This, in turn, has provided license to all ecologists to frequently conclude, and not infrequently assert, that their particular hierarchical level of study, be it physiology, demography, or rainforests is most relevant to the particular issue at hand. Advancing technologies have furthered this licensing process. For example, as soon as computers became readily accessible and the solution of even small matrix algebra (i.e., sets of linear algebraic statements) problems became feasible, development of the Leslie Matrix became the standard by which population analysts and demographers conducted discrete cohort demographic analyses. Roughly the same matrix algebraic approach is now readily applicable to a much wider array of biological and evolutionary problems. And on and on. Once Geographical Information Systems (GIS) and Geographical Positioning Systems (GPS) technologies became readily available, former paradigms involving ecology and earth science began to crumble. Laypersons' concepts of earth and the human influence on its balance or lack thereof virtually changed overnight (at least by an old professors' time scale). The development of the discipline of landscape ecology now became inevitable. People, God forbid, simply must be considered, very explicitly, in any realistic and/or practical models of ecological systems.

A second reason was provided by Biosphere 2, the scientific endeavor to determine how much (or how little) we did in fact understand about balanced ecological systems. The designers of Biosphere 2 attempted to create a closed ecological system (external energy sources were used to fuel Biosphere 2) designed to support eight humans with air, food, and water for two years. Though invaluable results were obtained and are still being analyzed, Biosphere 2 failed the principal stated mission. According to Cohen and Tilman (1996) "Isolating small pieces of large biomes and juxtaposing them in [an] artificial enclosure changed their functioning and interactions rather than creating a small working Earth." Landscape ecologists have long recognized that the context of each ecosystem matters; some, such as Harris, believe that the contextual setting may actually be more important than tinkering with the "content," at least when considering landscape-ecological function and biodiversity conservation value of protected areas (which are notoriously too small). Not only was the high fraction of vertebrate extinctions in Biosphere 2 (19 of 25) not anticipated, all pollinators disappeared, and even most insect species went extinct. Areas that were designed to be deserts transformed into chaparral or grasslands. Cohen and Tilman concluded that "there is no demonstrated alternative to maintaining the viability of Earth.....Dismembering major biomes into small pieces, a consequence of widespread human activities, must be regarded with caution....Earth remains the only known home that can sustain life."

Are the implications of the Biosphere 2 experiment likely to have any immediate consequences for the human assault on Biosphere 1, namely the home planet? We doubt that the results of such ecological experiments will find their way into public policy any time soon. And this simple proposition elevates to yet higher levels of recognition for research, understanding, management, and policy decisions that are based on the ecology of landscapes. This "regional-scale" approach that can include numerous interactive ecosystems is most obviously essential to biodiversity conservation. No one knows how small a park or preserve can be and still function as a safe haven for species of different sizes or trophic or critical ecological processes under different levels of primary productivity.

Central Park in New York City contains a lake, forests, and open fields and, in fact, was formerly a "hot spot" for biodiversity. However, context does matter, for the formerly manifold connections to surrounding natural systems have been severed and much native biodiversity has been lost. As landscapes become fragmented, so too do certain transecosystem exchanges and cross-system ecological processes such as international migration of organisms, not to mention the nature of birds that depend not only upon cavities in trees for nesting, but the sallying forth in much more open spaces for feeding. No doubt, there are many reasons why landscape fragmentation continues unabated. But, to the extent that environmental scientists continue to focus on the homogeneity aspects of ecosystems (lake vs. marsh vs. tree island or forest), the tragedy will continue. One might even go further to assert that environmental scientists have become their own worst enemy when it comes to understanding and managing for the heterogeneity so necessary to the functioning of landscape level systems. Sax (1991) purposely arouses our sensitivities when he states that, "A fundamental purpose of the traditional system of property law has been to destroy the functioning of natural resource systems." (1991, xx). With precious little regard for genetic variation that only occurs across geographic regions, humans continue to isolate and then proceed to erode qualitative aspects of critical habitat.

Sadly, while the abundance of a few species that are either endangered or obligatorily tied to the former habitat decreases to less than viable levels, many species that are already common actually colonize and increase their numbers such that unsuspecting observers are beguiled into believing that fragmentation is good. Well, indeed it is, if one desires more of what is already common. Many state and national land acquisition programs are explicitly predicated upon the notion that fragments of habitat, whether surrounded by more naturalistic systems or by turnpikes and shopping malls, will function well into the future as biodiversity preserves. Based upon the slightly more than 100 years of experience with parks and protected areas it seems quite clear that such is not the case.

Conversely, purposeful introduction of exotic species proceeds unabated, if not at an accelerated rate, at face, not only to increase local diversity, but to be justified by more compelling demands such as short-term productivity. But, it is our contention that the maintenance of *in situ* biodiversity integrity simply cannot persist over the long-term in the absence of effective conservation of the underlying ecological processes that both nurtures and maintains the diversity in the first place. We might therefore ask what process is being preserved in a park or protected area. Setting aside a park or preserve or saving a "hot spot" seems a futile exercise indeed unless some critical process benefits. The interruption of ecological processes across landscapes disturbs us. When a fire or an organism fails to cross a human-created barrier, such as a road or agricultural field, we suggest that critical processes such as propagation, dispersal, and movement have been disrupted. Our mission here is to aid in the restoration of such processes.

In Part I, Chapter 1, we begin with a brief history of landscape ecology. In Chapter 2, we discuss epistemology, the study of how we know what we know. This forms the foundation to organize our presentation. We adopt a top-down approach to the study of the ecology of landscapes in Chapters 3 and 4. Landforms (landscapes without life) are discussed in Chapter 4. We then add the biota, keeping in focus the top-down effects of biotic processes.

We introduce the ecology of landscapes in Part II. We emphasize the difference between landscape effects and landscape ecology in Chapter 5. Lennart Hansson describes his vision of landscape ecology using examples from Sweden in Chapter 6. To be useful, theories must be put into practice, the objective of Part III. We present several landscape theories and discuss remembering fragmented landscapes in Chapter 7. Bob Ulanowicz in Chapter 8 discusses ecosystem ascendancy as applied to landscape ecology. Anyone traveling across America today cannot help but notice the great homogenization that has occurred from coast to coast. Turner and Rylander of the Conservation Fund present their analysis in Chapter 9. The creation of corridors in Europe is highlighted in Chapter 10 by Rob Jongman and Daniel Smith. Pijanowski, Gage, Long, and Cooper present results of a landscape change model developed to predict the future course of land change in Michigan in Chapter 11. A large effort to model ecological processes in the Everglades of Florida is described by DeAngelis, Gross, Wolff, Fleming, and Nott in Chapter 12. In the Everglades we see the full implication of external threats to an inherently heterogeneous landscape in both space and time containing several threatened and endangered species juxtaposed between a large, expanding metropolitan city and an encroaching sea.

We accept that some of our colleagues will take issue with some of our views. For instance, more than one of our colleagues has suggested that landscapes depend upon the organisms viewing them. While we believe this is true, we feel that the conservation challenges we face today require a more anthropocentric view. Because we are interested in the ecology of landscapes (viz., landscape ecology in the vernacular), we must first exclude the patches of topiary from British gardens and certain of those in Hollywood so that we can focus attention on the much larger expanses of land and co-evolved biotic systems that occur thereon. These expanses would usually transcend over hundreds of square kilometers, but such would not always be the case because we have looked up valleys that clearly contained two or more different ecosystems and the entire area transcended but tens of square kilometers. Moreover, because we care about the predominance of landscapes on earth (as opposed to seascapes, which occupy 66%), they would of necessity involve life. Life occurs in sand, it occurs on beaches, on beach dunes, in deserts and even on ice-clad mountain peaks. Therefore, it does not seem unreasonable that the landscapes we refer to herein contain life.

Given that a landscape contains life (as opposed to moonscapes or Venus-scapes), it is then important to wonder about the relation between a landscape and the life that it contains. For that matter, it is, in our judgment, important to wonder about the role that life plays in the landscape itself. We assert that any landscape under question or scrutiny would not be the same if it had not previously been, or is presently under, significant influence of life. And so, for purposes of moving the discussion forward, we assert that the only landscape that is important in the field of landscape ecology contains life and is in some sense influenced by that life. Perhaps there are scapes, and perhaps they are made of land, that do not contain life or that are not influenced by life. But we do not deal with them here.

A convergence of themes in ecology is rapidly occurring, lending firm support to the study of the ecology of landscapes. So-called "ecosystem engineers" have renewed interest in how certain species interact with their environment (Jones et al. 1994). With only a short step the importance of mobile organisms that help create, maintain, and exploit more than one ecosystem can be imagined. Also, Wilson's extension of multilevel selection theory that seeks to explain community-level selection in local communities brings an evolutionary approach to the study of organisms across a mosaic of habitats (Wilson 1997). A research program integrating these two powerful themes on landscapes is now possible. We believe that advances in the study of landscape ecology will come from studies in landscape effects (the effects of pattern on process), and mobile organism's top-down effects on landscapes (the effects of mobile organisms on pattern). Heretofore, the study of landscape effects has preoccupied landscape ecologists. The study of the ecology of ecosystem engineers and multispecies selection will lead naturally to the study of other organisms whose top-down effects create and maintain landscapes, completing the circle necessary to firmly establish the discipline of landscape ecology within an evolutionary context.

However, as taught in basic anthropology, class structure is a highly pervasive phenomenon among humans and other higher vertebrates, and one certainly exists in science, in general, and in ecology specifically. Thus, new ideas or concepts are not evaluated just on their merits, but rather, who does the evaluation of whom and what. Landscape ecology is at this stage of growth and development now; wonderfully gifted ecologists who have practiced at many different scales are sincerely (for the most part) and forcefully debating terms, concepts, approaches, and results. This is good.

Editors

Jim Sanderson is currently working for Los Alamos National Laboratory in New Mexico. He is also Director of Conservation for Mountain View Farms and Conservation Breeding Centre, Vancouver, Canada. He received his Ph.D. from the University of New Mexico in 1976. An avid traveler, Dr. Sanderson has collected and synthesized wildlife issues from around the world. His interests include quantitative ecology, community ecology, and landscape ecology. He maintains an active research program on small wild cats in South America.

Larry D. Harris is Professor Emeritus at the University of Florida. He began his professional career as a conservationist trained in the Midwest and then east Africa. After finishing his Ph.D. in systems ecology, he worked in the systems modeling group for the grassland biome project of the international biological program (USIBP). He has spent the last 27 years developing biodiversity conservation programming at the University of Florida, as well as being a globally active consultant on this subject. His work that has received the most attention and award recognition concerns the role and importance of forests to the perpetuation of biodiversity on earth: in Florida as well as globally.

Contributors

E. Jane Comiskey Department of Ecology and Evoluntionary Biology, University of Tennessee, Knoxville, Tennessee

William E. Cooper Institute for Environmental Toxicology and Department of Zoology, Michigan State University, East Lansing, Michigan

Donald L. DeAngelis U.S. Geological Survey, Biological Resources Division, Department of Biology, University of Miami, Coral Gables, Florida

D. Martin Fleming U.S. Geological Survey, Biological Resources Division, Everglades National Park, Homestead, Florida

Stuart H. Gage Spatial Analysis Laboratory, Department of Entomology, Michigan State University, East Lansing, Michigan

Louis J. Gross Department of Mathematics, The University of Tennessee, Knoxville, Tennessee

Lennart Hansson Department of Conservation Biology, SLU, Uppsala, Sweden

Larry D. Harris Professor Emeritus, University of Florida, Gainesville, Florida

Rob H.G. Jongman Wageningen Agricultural University, Department of Environmental Sciences, Land Use Planning Group, Wageningen, The Netherlands

David T. Long Geochemical and Isotope Laboratory, Department of Geological Sciences, Michigan State University, East Lansing, Michigan

M. Philip Nott Department of Ecology and Evolutionary Biology, The University of Tennessee, Knoxville, Tennessee

Bryan C. Pijanowski Spatial Analysis Laboratory, Department of Entomology, Michigan State University, East Lansing, Michigan

Jason Rylander The Conservation Fund, Arlington, Virginia

James Sanderson Los Alamos Laboratory, Los Alamos, New Mexico

Daniel Smith Department of Wildlife Ecology and Conservation, University of Florida, Gainesville, Florida

John F. Turner President, The Conservation Fund, Arlington, Virginia

Robert E. Ulanowicz University of Maryland, Chesapeake Biological Laboratory, Solomons, Maryland

Wilfried F. Wolff Department of Biology, University of Miami, Coral Gables, Florida

Contents

Part I

The Presence of the Past

1

Brief History of Landscape Ecology

Jim Sanderson and Larry D. Harris

CONTENTS

The 1949 publication of what is affectionately referred to as "The Great APPES" (i.e., Principles of Animal Ecology, by Allee et al.), the close follow-up by Adrewartha and Birch (1954), as well as the magnum opus of field zoo-geographer, Darlington (1957), brought the era of descriptive field ecology and matters of distribution and abundance of species to an honorable close, at least in The United States. The era of laboratory, experimental, and mathe-matical ecology quickly filled any existing niche space as by 1962 Preston (1962a, 1962b) and the immediacy of MacArthur and Wilson's seminal work in 1963 initiated a new era for ecological thinking. Those of us sufficiently old to remember the popular press can wax "oh' so" eloquently about how 'ecol-ogy had now come of age.' It is easily arguable that there had been a major paradigm shift in Kuhn's (1962) sense of the concept.

A second happening involved the formal establishment of expensive, large-scale investigations under the aegis of The International Biome Pro-gram (IBP). These large scale, but all too descriptive research programs are epitomized in North America by the Grassland Biome Program centered at Ft. Collins, Colorado, and the Eastern Deciduous Forest Biome program cen-tered at Oak Ridge National Laboratory in Tennessee (Golley 1993).

Although now the subject of too much ridicule, these grand and formal-ized, research programs led quickly to establishment of three Systems Ecol-ogy capacity-building grants that married the disciplines of mathematics, systems science, and ecology. Professor Frederick Smith, then at the Univer-sity of Michigan, was not only central to the transformation just described, but he made yet one other, perhaps final, tactical maneuver. The world renowned Harvard Graduate School of Design and appropriate administrators recognized the need to bring formal ecological thinking into landscape architecture and regional planning programs; Professor Smith accepted stewardship of what was arguably the first official landscape ecology chair at a major university in the U.S. Other universities quickly followed in spirit, if not quite as formally.

At the Oak Ridge National Laboratory, the transition from systems ecology, as it had been conceived and executed in the Biome program transformed rather seamlessly into initiatives in what is now referred to as landscape ecology. Indeed, the explicit statements of this (as occurring in grant proposals and personnel recruitment) became obvious by the early 1970s. Their programming efforts led to a compilation of publications (Burgess and Sharpe 1981) that effectively melded the great descriptive data bases of the first half century of forest ecology in the eastern U.S. with the ongoing, but now dwindling, U.S. IBP research programs with the IBT paradigm which had captured a lot of attention by the mid to late 1970s. Even though descriptive field ecology found or created effective new niches (e.g., Organization for Tropical Studies, OTS), primarily in the tropics, the TIB and/or the 'patch-in-a-matrix' concept (Islands in the Stream?) had seemingly captured the budding research programs in landscape-level ecology.

Continental reserves surrounded by unnatural landscapes, for instance, became island reserves (Diamond 1975; Diamond and May 1976). Diamond (1975) argued that island biogeography results could be applied to the design of landlocked forested nature reserves and isolated mountain tops, so-called virtual islands (Diamond and May 1976). Maximum area and minimum perimeter, Diamond suggested, were critically important variables in reserve design, and the restoration of corridors that interconnected now dismembered landscapes, he argued (as Preston had said in 1962), could act to increase species persistence by increasing the effective area. Some of the successors of certain U.S. IBP groups pursued the tangent of watersheds as the next most promising endeavor. They chuckled when asked why not study wolf sheds, and it is reasonable to conclude that they had missed the point. But most importantly, the ecosystem, paradigm, and all that it stood for persisted in guiding conventional ecological research. A branch of ecologists, we refer to as community ecologists, aggregated with other sympathetic forces of biodiversity conservation to form a new organization: The Society for Conservation Biology. Needless to say, the ecosystem paradigm necessitated a focus on energy flow, trophic food webs, nutrient cycles, etc. Interactions between components of the systems were investigated almost mechanically, and output variables such as productivity, measured in units of $gm/m2/yr$, became a currency of ecosystem ecology. Although interactions within ecosystems were studied through time, major advances were made toward linking the biotic and abiotic components via the soil, to leaf, herbivore, carnivore, and decomposition (spatial heterogeneity was largely relegated to a different agenda). The study of species and the physical and chemical processes of their environment could now be taught like any other engineering discipline cookbook style. Though Golley as Division Director of Environmental Biology at the National Science Foundation from 1979 to 1981 asserted that humans be included in these study systems (as they were in Europe), momentum carried the ecosystem paradigm forward without them.

While debates such as those on reserve design (Single Large vs. Several Small — SLOSS) diverted attention from more pressing issues (Diamond 1975; Terborgh 1976; Simberloff and Abele 1976), Kushlan (1979), working in Everglades National Park in Florida, argued that the Theory of Island Biogeography did not quite apply as was assumed. The shifting pattern of population changes in 16 species of ciconiiform wading birds species indicated that the application of island biogeographic theory to the design and management of continental wildlife reserves required more consideration. Isolation of a continental reserve could lead to ecosystem degeneration, the extent and rapidity of which depended on the ecological condition of *adjacent* habitat. Here we find the profound significance of Kushlan's results — the recognition that the contents of a protected area could be negatively impacted by the contextual setting of the area. Conflicts between species management and ecosystem management illustrated the need for a regional basis for preservation.

Kushlan (1979) realized that size alone was an inadequate measure of the effectiveness of a reserve. Everglades National Park was 5670 km^2; it was bounded by Big Cypress National Preserve of 2370 km^2 and three Water Conservation Areas totaling 3490 km^2, making the total protected area about 12,000 km^2. The importance of environmental heterogeneity and maintenance of the functional characteristics of the reserve, such as the timing of changes in water levels beyond the park boundary, had to be considered. Spatial isolation from the buffering of contiguous habitats had resulted in quantitative and qualitative alteration of the functional relations within the reserve that led to environmental degradation and the decline in wading bird populations. Because local extirpations might occur in highly specialized species, Kushlan recommended a regional approach to the management and perpetuation of biodiversity that would permit recolonization from refugia when conditions changed. Environmental heterogeneity at the scale of the landscape was critical to maintaining biodiversity, especially in managed landscapes.

Crisis in Conservation

In 1980, Soulé and Wilcox sounded the alarm in the U.S. Whatever ecologists were doing was not working. In 1973, the 95th U.S. Congress amended the Endangered Species Act establishing, among other things, a legal mechanism known as "taking" for causing harm to a protected species. Ehrlich (1980), in the final chapter of Soulé and Wilcox (1980), claimed that the momentum of human exploitation of natural resources was likely to overwhelm the biosphere. He warned that for every hard-won battle, the forces of conservation "suffer crushing, if unheralded, defeats as unknown populations and species are plowed under from Anaheim to the Amazon." Unless the trends of the past were suddenly and decisively reversed, conservationists could only hope to "slightly delay an unhappy end to the biotic Armageddon now underway." Ehrlich and Ehrlich followed in 1981 with their book, *Extinction*.

Because years of field research and data collection were necessary to produce valuable results in ecological studies, the momentum built into scientific inquiry could not suddenly be terminated and redirected. Lovejoy et al. (1983, 1984) studied isolated forest tracts in Amazonia and argued that the results of MacArthur and Wilson applied. Harris (1984) wrote of fragmented forests in the northwestern U.S. and referred to island biogeographic theory in his book, *The Fragmented Forrest*. That terrestrial reserves were not habitat islands was clear to these authors. However, Harris wrote that "the whole module should be programmed into the context of a production-oriented landscape. This allows the preservation areas to be buffered from the harsh impacts and vicissitudes of the human-dominated landscape."

Europeans, quite independently from the U.S., were pursuing their interests in the ecology of landscapes. The Netherlands Society for Landscape Ecology (NSLE) organized a conference in Veldhoven, The Netherlands, on April 6–11, 1981 (Tjallingii and de Veer 1982). NSLE was founded in 1972 "to gain a deeper understanding of the structure and functioning of landscapes and the patterns and processes in landscapes" (Wijnhoven 1982). Americans Julian Fabos, Richard Forman, Frank Golley, and Richard Sharpe attended. Through a series of lectures, workshops, and posters, Europeans presented their vision of landscape ecology. Most presentations addressed the negative impacts humans had upon the European landscape. Though aesthetics and architecture were integral components of European landscape ecology, van der Maarel (1982) wrote of the far-reaching side effects that humans had on nature reserves. "This makes nature reserves rather different from islands in the sea. Thus from a landscape-ecological point-of-view we must again further modify the theory of island biogeography." He suggested that "landscape ecological theory" should play a major role in planning nature reserves.

Forman spent 1982 with Godron at the Centre d'Etudes Phytosociologiques et Ecologiques L. Embarger in Montpellier, France. At the Veldhoven conference, Forman (1982) presented his preliminary vision of landscape ecology. There he espoused the necessity of a contextual analysis of landscapes. A landscape, Forman suggested, was a matrix with patches and corridors where interactions occurred. Though he did not use the word "context" he referred to "specific linkages that exist with surrounding landscape elements" that must be considered when "making land-use decisions." Theme IV of Veldhoven was devoted to the conservation of natural areas. The species, reserve, and resource-oriented approaches to conservation were all discussed, yet the work of Soulé and Wilcox was not referenced by a single speaker. In his closing remarks, Zonneveld as Chairman of the Congress Organizing Committee stressed the importance of the formation of an international society for landscape ecology. A year later Naveh (1982) explained that landscape ecology had gained a general recognition as a branch of modern ecology in central and eastern Europe and Israel. He used as examples the chairs in landscape ecology at several universities in Germany. "The English-speaking world, and especially the United States, is almost totally unaware of these developments," he wrote. He attempted to spell out the theory and general principles Forman sought.

While papers and books fueled by the Theory of Island Biogeography continued to pile up and IBP research results flooded the American ecological literature, Forman convinced Paul Risser, then Chief of the Illinois Natural History Survey, and ecologist Jim Karr to host a meeting to discuss a new approach to research in ecology. Neither Risser nor Karr had experience in landscape ecology. The meeting was by invitation only and was strongly influenced by the IBP ecologists because no one in attendance, aside from Forman, Godron, and Golley, had more than a cursory knowledge of landscape ecology. The now historic meeting took place at Allerton, Illinois.

Allerton Park

On April 25–27, 1983, 25 attendees met at Allerton Park to discuss the foundation of a new synthetic discipline referred to as "regional ecology" or "landscape ecology." Previous attempts to achieve a synthesis had failed, they claimed. A persistent nagging recognition prevailed throughout the meeting — either a new discipline or area of specialization would emerge alive and vibrant or be stillborn and forgotten. Given the historical background, Forman and Golley were probably determined to push through a "new" science of landscape ecology. The problem was convincing the rest of the invitees that this was the right thing to do. The ideas discussed were not new and had been presented in the European literature over the preceding decade. The time had arrived to collectively discuss landscape perspectives in basic and applied research on natural resources that, according to Risser et al. (1984), were "stalled by several converging themes" such as:

1. a preoccupation with the extension of island biogeography theory to continental landscape patches;
2. the presumption that ecosystem-level characteristics were adequate to address landscape-level characteristics;
3. a recognition of the need to address landscape issues in land and resource management;

4. a belief that map-overlay methodology was sufficient to capture the essential attributes of multiunit landscapes;

5. the realization that human activities were an integral part of any meaningful concept of landscape ecology; and

6. the recognition that the inclusion of many appropriate scientific disciplines results in an exceedingly complex field.

Although a landscape perspective in ecology was not new (Leopold 1949; Neff 1967; Troll 1968; Naveh 1982; Tjallingii and de Veer 1982), a firm theoretical basis for an ecology of landscapes was missing. Several authors (Forman 1982; Hansson 1977; Naveh 1982; Naveh and Lieberman 1984) were attempting to generalize ecology to guide research management, but without a definitive, ecologically based theory and methodology how could natural resources be managed?

Attendees "agreed" that landscape ecology should consider the development and dynamics of spatial heterogeneity, spatial and temporal interactions and exchanges across heterogeneous landscapes, influences of spatial heterogeneity on biotic and abiotic processes, and management of spatial heterogeneity. In 1984, the primary focus of landscape ecology was on:

1. spatially heterogeneous areas such as pine barrens (Forman 1979) and regions of row crop agriculture, Mediterranean woodland landscapes, and areas of urban and suburban landscapes;

2. fluxes or redistribution among landscape elements; and

3. human actions as responses to, and influences on, ecological processes.

The relationship between spatial pattern and ecological processes was not restricted to a particular scale. For instance, Weins (1985) discussed how organisms reacted to patterns in the environment. Landscape heterogeneity had previously been recognized as being of fundamental importance in landscapes (Whittaker and Levin 1977). Interactions at different scales were thought to have varying effects. Although hierarchical approaches offered a structure for organizing thoughts (Allen and Starr 1982), a necessarily hierarchical structure was not endorsed, though no other organizing principles were presented.

"Fundamental questions" were raised by Allerton Park attendees that addressed the development, maintenance, and effects of temporal and spatial heterogeneity of the landscape:

1. How were fluxes of organisms, of material, and of energy related to landscape heterogeneity?

2. What formative processes, both historical and present, were responsible for the existing pattern in a landscape?

3. How did landscape heterogeneity affect the spread of disturbance?

4. How could natural resource management be enhanced by adopting a landscape ecological approach?

Missing were keywords such as content, context, and juxtaposition. References to European works were sparse, largely because they were unknown. Kushlan's work was not cited. Only brief mention of the effects organisms had on maintaining or extending their environments was made. Conservation biology was not mentioned specifically. The recognition that no unifying theory had developed was testimony to the uncertain future, at least in the U.S., of what we now call landscape ecology. Only one European, Godron, attended Allerton Park. The importance of the flow of energy and materials through ecosystems placed on landscape ecology was testimony to the momentum of the IBP program and the built-in biases of most of the participants. The ecology of landscapes was almost stillborn in the U.S.

The Eternal External Threat

Kushlan's (1979) observation that landscapes and the contextual setting of parks and reserves were important was cited less than a dozen times in the ensuing seven years. Nevertheless, his conclusion that context was important gained converts. Janzen (1983) fully appreciated the call of conservationists and wrote that small islands of reserves were only poorly analogous to more conventional islands surrounded by water. Three years later Janzen (1986) warned of "the eternal external threat." Here we see again the connection between conservation biology and landscape ecology powerfully spelled out with clear examples of why the ecology of landscapes differed from island biogeography. Forman and Godron published their popular book, *Landscape Ecology,* in 1986 without citing Kushlan's or Janzen's articles, but included references to Soulé and Wilcox (1980).

Another approach to the study of fragmented landscapes had been developing in Canada under Merriam (1984), also an attendee at Allerton Park (Middleton and Merriam 1981; Fahrig 1983; Fahrig and Merriam 1985; Middleton and Merriam 1983). Based on the study of metapopulations (Levins 1969, 1970), Merriam's approach was to study small mammals in farm field fragments connected by corridors of favorable habitat. Animals needed to move through the landscape in response to changing resources needs. The frequency of extinctions in patches, Merriam found, depended on their degree of isolation from other favorable patches.

Three Allerton Park attendees, Urban, O'Neill, and Shugart (1987), as if viewing a van Gogh, wrote that the science of landscape ecology was motivated by the need to understand the development of pattern in ecological phenomena. Terrestrial landscapes, they observed, consisted of heterogeneous land forms, vegetation types, and land uses whose development and dynamics required attention by ecologists. Pattern was the "hallmark of a landscape" and they presented a hierarchical paradigm of landscape ecology generalized from their experience with forested landscapes.

Landscapes, Urban et al. (1987) suggested, were mosaics of patches created by disturbances, biotic processes, and environmental constraints acting across varying temporal and spatial scales. The authors' spatial scale vs. temporal scale graphs of disturbance regimes, forest processes, environmental constraints, and vegetation patterns are seen today in a variety of contexts. The loosely coupled, multilevel organization of landscapes required a hierarchical theory to adequately address complexities (King 1997; O'Neill et al. 1986; Allen and Starr 1982). The slicing-and-dicing approach to landscape quantification began and continues to advance today (Hargis et al. 1997).

Components of a hierarchical structure were organized into levels according to their functional scale. Events at a given level have a characteristic natural frequency and a corresponding spatial scale. Low-level events were viewed as being comparatively small and fast, and higher-level events were large and slow. By presenting an example derived from eastern deciduous forests, Urban et al. (1987) generalized their results to other landscapes. Four levels of a forest hierarchy led to the definition of a landscape. Forest gap creation took place rapidly over a small spatial scale, e.g., a stand of trees might have several gaps and disturbance-created patches; a watershed consisted of local drainage basins and topographic divides; and a landscape might be multiple watersheds with different disturbance regimes and be influenced by different land use practices. Today, use of the term 'level' is discouraged (King 1997).

The recognition that human activities could influence the landscape led Urban et al. (1987) to include human impacts on ecological processes. Anthropogenic effects often rescaled patterns in space and time and acted to homogenize patterns through land use practices and monotypic species introductions. Such activities were seen capable of causing local as well as regional declines in certain forest microhabitat specialists.

The purpose of a paradigm is to organize thinking and offer a conceptual and analytic framework for future work. The hierarchical theory presented by Urban et al. (1987) fit well with approaches to problem solving in general. Many human systems were organized hierarchically such as our present political and military systems. Even the species concept is organized hierarchically. However, other human societies often organize thinking differently. While Western thinking is typically hierarchical, Eastern thinking is often dualistic, with the active, masculine yang element or force balanced by an opposite yin, the passive female element or force. In any case, Western thought now favored the study of some form of hierarchically based landscape ecology (Zonneveld 1988). But do landscape patterns imply functional organization?

While ecologists tried to organize their thinking, the fragmentation of natural habitats was becoming increasingly more important in conservation biology. The results of habitat fragmentation were (1) habitat loss and (2) habitat insularization, both of which contributed to the decline of biological diversity (Wilcox and Murphy 1985; Miller and Harris 1977). The Theory of Island Biogeography was once again invoked to quantify the loss of species. Demographic stochasticity, environmental variation, genetic stochasticity, and natural catastrophes led species to become extinct (Shaffer 1981). Fragmentation was seen as increasing the likelihood that one or more of these events would indeed occur.

The Bay checkerspot butterfly *(Euphydryas editha bayensis)* that had been studied for many years by Murphy, Wilcox, and colleagues served as an example. The butterfly survived in areas of north-facing sites under dry conditions and in sites with southern exposure during wet years. A variety of heterogeneous sites was thus critical to the survival of the butterfly as was connectivity between sites. Little did the experts realize that in 1997 they would document the extirpation of the butterfly. A combination of bad weather and lack of food plants led to the butterfly's demise.

In 1983, Noss (1983) suggested that landscapes should replace ecosystems as the unit of management. Dynamic natural biotic and abiotic processes had to be maintained over much larger areas and these areas might need to be connected by a network of corridors. The negative effect of edges (Gates and Gysel 1978) and the recognition that many species required several different ecosystems to exist demanded an expanded view of conservation. Rejuvenating natural phenomena such as fires might destroy small reserves, whereas large natural areas depended on fires for their existence and perpetuation. Furthermore, species composition and abundance and not number were an important metric for conservation. Moreover, the size of the preserve should be made simply proportional to the magnitude of a typical natural episodic event (Noss 1983).

Noss and Harris (1986), with reference to the Theory of Island Biogeography, also realized that spatial dynamics were an important and overlooked process in conservation efforts. They wrote that comprehensive planning at the scale of the landscape was essential to maintaining ecological processes. Integration of reserves, they suggested, was critical. According to Noss and Harris "An expanded focus from just within-boundary conditions (*content*) to the regional (*context*) is necessary if our intent is to mitigate the external pressures that impinge on protected areas." Mitigation of habitat fragmentation was essential to preserving "spatiotemporal heterogeneity and landscape interactions." Noss and Harris realized that processes, not necessarily the products of these processes, should be preserved.

Scientific paradigms die hard, however, and the late 1980s saw conservationists continuing to embrace the Theory of Island Biogeography to argue their case. Newmark (1987) used a land-bridge island analogy to explain the loss of mammalian species in 14 large western parks in North America. The total number of extinctions exceeded the total number of colonizations within the reserves, the number of extinctions in the reserves was inversely proportional to their areas, and the number of extinctions was related to reserve age, just as the theory had predicted. Newmark recommended that the parks be enlarged or that lands bordering the parks be cooperatively managed to avoid future extinctions. In either case, Newmark's message was clear. While some scientists argued with Newmark's methods, those with a better appreciation of landscape ecology could have predicted his results with no help from the Theory of Island Biogeography crutch. Newmark cited neither Kushlan (1979) nor Janzen (1983, 1986).

In 1987, the U.S. International Association of Landscape Ecology (U.S. IALE) was established, *Landscape Ecology* became the journal of the association, and Golley (1987) assumed a leadership role. The 1990s have seen an explosion of work in landscape ecology. By 1991, metapopulation models had largely replaced the Theory of Island Biogeography as an organizing principle for the study of organisms in fragmented landscapes (Merriam 1991). Much work has been done by many of the attendees at Allerton Park. Nearly 300 scientists attended the 1997 U.S. IALE meeting held at Duke University. More than 600 scientists attended the international meeting in Snowmass, Colorado, in 1999. Where are we today?

Attendees at Allerton Park expressed hope that "the development of a specific theory that addresses issues of landscape heterogeneity will be expedited by collecting and analyzing empirical data, using model simulations, and searching for similarities in related disciplines from which to extract and formalize theory." This has not occurred. Indeed, the distinction of landscape and ecosystem is today disappearing as was evidenced by the announcement of the new journal *Ecosystems* at the 1997 IALE conference. The journal sought articles that dealt with ecological problems with spatial scales from bounded ecosystems to the earth, and time scales from seconds to millennia. Can the guiding paradigms of ecosystem studies be extended to include the study of landscape ecology?

Pickett et al. (1994) discussed the integration of ecology and presented ecology as a continuum of subdisciplines along a gradient dealing with strictly physical phenomena to strictly biological concerns and spanning the space from meteorology, geology, and hydrology to systematics, genetics, and physiology. Cherrett (1988) surveyed British Ecological Society members, asking them to rank the most important concepts in ecology. Ecosystems ranked first, followed by succession. Landscape ecology was not ranked among the top 30 concepts. Weins (1992) summarized research published in the first five volumes of the journal *Landscape Ecology*. He concluded that landscape ecology was a nonquantitative discipline concerned with broadscale features of land use (hectares to square kilometers) and human-induced landscape structure. Though strongly relevant to the study of habitat fragmentation, reserve design, the maintenance of biological diversity, and natural resource management, landscape ecology lacked a solid theoretical basis. Must landscape ecology remain heuristic or can theories be created to allow landscape ecologists to make predictions that can be tested?

With respect to biodiversity conservation, there was widespread concern that something was wrong, that something was missing. While Habitat Conservation Plans mandated by the Endangered Species Act were debated, private landowners asserted their rights, scientists continued to pile up interesting and unusual facts about organisms, and more and more species were being added to the list of contenders for threatened or endangered status. How many more studies and experiments were necessary? Just as in nearly every public election in the U.S., consensus was never 100%. Must all scientists, not just those familiar with the problems, be in complete agreement? Perhaps a simple majority opinion should have been considered sufficient for action.

Is there a difference between the ecology of ecosystems and that of landscapes? Are there collective and emergent properties of two or more juxtaposed ecosystems? What is the difference between landscape effects and landscape ecology? Is there a difference between landscape metrics and landscape ecology? Can common concepts arising from studies of many different landscapes be used to create a theory of landscape ecology? Must this theory in turn then apply to every landscape? Must it apply to any landscape? Are Habitat Conservation Plans as specified by the Endangered Species Act a way to protect species? Is the "new" paradigm of Ecosystem Management (Grumbine 1994a,b) the final word in managing our natural resources and conserving biodiversity? We will explore these issues and others in detail by adopting a top-down approach to the study of landscape ecology.

A Top-Down Approach

Our approach to landscape ecology is to emphasize the effect mobile organisms have on the functioning of ecosystems. We believe an understanding of landscape ecology cannot be achieved by the paradigm of bottom-up thinking as espoused by a half-century of IBP ecosystem studies. We agree with Bissonette (1997) that to understand processes at level L one must approach the problem from level $L + 1$. We argue that context is equally as important as content and that an isolated, dismembered landscape fragment will inevitably lose natural biodiversity. Fragmentation of the greater landscape continues and even the largest western national parks are "relaxing," a euphemism for losing species (Newmark 1995). We further believe that the most detailed mathematical models of all the biodiversity within a landscape (proximate models) will not suffice to predict the outcome of management practices if the contextual analysis reveals that human impacts outside the landscape (ultimate causes) are contributing to the untimely demise of a reserve. From our vantage point we will demonstrate that protecting disconnected vignettes of nature in isolated national parks and reserves or saving so-called "hot spots" of biodiversity as has been recently espoused (Dobson et al. 1997) simply will not work.

The bottom-up study of ecology popularized by the IBP program continues to provide valuable insights into ecological processes. The discipline of macroecology (Brown 1995) seeks to explain the distribution and abundance of organisms. However, instead of asking what factors determine the distribution and abundance of organisms, a top-down approach to the study of landscape ecology seeks to explain how the distribution and abundance of organisms affects the entire collection of biodiversity and processes on the landscape.

Many authors have discussed landscape-level effects. For instance, the effects of landscape fragmentation on birds has been described (Donovan et al. 1997; Merriam and Wegner 1992; Terborgh 1989; Wilcove 1985; Noss 1983; Whitcomb et al. 1981). While landscape effects are important, the top-down effects that mobile organisms have on structuring and maintaining landscapes have been sorely neglected (Jones et al. 1994, Naiman 1988). The activities of beavers, deer, elephants, bears, wading bird colonies, and a few seed-dispersing organisms have been documented. However, hundreds of millions of years of evolution lead us to believe that all organisms have equally important, but perhaps less obvious impacts upon the landscape. Indeed, conservationists accept that all species are worth saving if only for ethical reasons. We submit that all species are worth saving in their natural environment because their activities are the essential processes that support and maintain those environments. We agree with Simberloff (1998) and Hansson (1997) that understanding the role of keystone and indicator species, respectively, to elucidate their roles in ecosystem functions and structuring processes is important. Understanding these roles in the context of heterogeneous landscapes is yet more important.

Multilevel selection theory has been used to explain the functional organization of individuals and single-species social groups. Wilson (1997) argued that multispecies assemblages can acquire properties associated with single organisms. That is, individuals are members of local communities, and are subject to selective forces operating within the community. Goodnight (1990a,b) showed that a two-species community of flour beetles (*Tribolium castaneum* and *T. confusum*) responded to selection as a single interactive system because the community was the object of selection. Indeed, Wilson (1997) pointed out that:

> The same evolutionary forces that produced the extreme functional integration that we call organisms may have produced more moderate functional integration in other multispecies assemblages that currently is not recognized at all.

Selective pressures are the integration of all processes in the local community. Leigh (1991, 1994) analyzed the structure of tropical forest communities in terms of multilevel selection theory. Previously, Odum (1969) argued that microecosystems were functionally organized, but gave no structuring mechanism. Dawkins (1982) has argued otherwise, however.

Biotic processes exert a strong influence on the atmosphere (Malin 1997; Gage et al. 1997), and the Gaia hypothesis argued further that the biota acted as a single organism capable of regulating processes on a planetary scale. Such effects do not imply functional organization, however. The emergence of patterns in the landscape also does not imply functional organization. Natural selection requires variation among individuals to operate. Complex interactions create variation in individuals and so enables selection to operate.

Wilson (1997) concluded:

> The evolution of species can be influenced by other species in the community, but it remains a separate entity with its own survival strategy of survival and reproduction. When natural selection operates at the community level, all of the species in a local community become part of a single interacting system that produces a common phenotype, more like genes than species as we usually think of them, and the local community acquires the properties of adaptation that we usually associate with individuals.

Therefore, while a beaver might very well be a keystone species in a mountain stream in New Mexico and in Maine, the genetic diversity contained in the beaver is likely tied closely to its local community. Protecting within-species genetic reservoirs is worthy of our attention, but only when viewed from the perspective of the landscape. Disassociating beavers from their local communities and interbreeding them as in a zoo destroys biodiversity and ultimately is damaging to the species, hampering conservation efforts if they are needed.

Jones et al. (1994, 1997) defined physical ecosystem engineers:

> Physical ecosystem engineers are organisms that directly or indirectly control the availability of resources to other organisms by causing physical state changes in biotic and abiotic materials. Physical ecosystem engineering by organisms is the physical modification, maintenance, or creation of habitats. The ecological effects of engineering on other species occur because the physical state changes directly or indirectly control resources used by these other species.

Note that ecosystem engineers control flows of energy and resources, but do not necessarily participate in these flows. Therefore, food web characterizations and ecosystem theory will not generally be useful for studying the effects of engineers (Jones et al. 1997). The study of ecosystem engineers is not new (Naiman 1988), but has only recently been revived (Jones et al. 1997). At the landscape scale, engineers modify and create mosaics of habitat and therefore enhance species richness and biodiversity, probably over evolutionary time as well.

Behavioral aspects often determine whether organisms are ecosystem engineers. Thus, attempts to predict *a priori* peculiarities of species in even simple communities will likely be unproductive. Jones et al. (1997) suggested that:

> ...there may be no substitute for starting with natural history and behavior in order to discover the key design feature and thereby understand the potential engineering impact.

Certainly, large organisms such as elephants, corals, and kelp are known to be engineers, but because the impacts of the organisms on their habitat have been so poorly studied, many more important engineers are likely to be identified. Indeed, we believe that many, if not most, top-down contributions of organisms to their environment are inadequately documented. Keystone species and ecosystem engineers are not synonymous (Jones et al. 1997), but at least confer special status to organisms so labeled. Highly mobile organisms, such as elephants, wide-ranging predators such as wolves, and migratory herbivores such as wildebeest might better be referred to as landscape engineers, for their activities are played out at the landscape scale encompassing more than one ecosystem and their impacts are noticeable even from space.

The paradigm of landscape ecology must rest soundly on theory. These theories enable the creation of hypotheses that can be tested. Our goal is to describe these theories and then demonstrate how these theories are applied to re-membering and restoring our native landscapes to improve the lives of humans that live within these landscapes. We will not be satisfied with saving mere remnants of the past. This can, after all, be done in zoos, botanical gardens, test tubes, and in virtual reality.

Conservation of Biotic Processes

Because humans are organisms, ecologists have naturally placed more emphasis on the study of organisms than on the processes that gave rise to and maintain the full spectrum of biodiversity that surrounds us. We must change our collective perspective and instead embrace the conservation of the biotic and abiotic processes that once occurred ubiquitously across landscapes. Our goal will be the restoration of natural ecological processes across the landscape that are perpetuated by living organisms and not to maintain certain landscape patterns. We will neither discuss measuring and recording landscape metrics nor present the use of geographic information systems. Though these are important quantification tools they are detailed elsewhere (Frohn 1997). To achieve our goal we must adopt a top-down approach to the study of the ecology of landscapes. The formation and maintenance of heterogeneity and variability of the landscape must be embraced. In the words of den Boer (1981),

> ...heterogeneity and changeability must be recognized as fundamental features, not only of the natural environment of a population but even of

life itself. The enormous genetical [sic] and phenotypical variation of a natural population is in some way a reflection of the heterogeneous and variable conditions in spite of which... it is able to survive for a shorter or longer time. And on a larger scale, the incredible diversity of life reflects the nearly infinite heterogeneity of natural habitats, which is again importantly increased by the presence and the actions of living creatures themselves. As long as heterogeneity and variability are considered to be mere deviations from "typical" cases, that are the only ones that are grasped by our intellect and caught in preconceived and often static (equilibria) theoretical structures, I fear we will deny some fundamental features of organic life.

On February 15, 1997, Stanford ecologist Harold Mooney, Secretary General of the International Council of Scientific Unions and a leader in the study of biodiversity, addressed the annual meeting of the American Association for the Advancement of Science in Seattle. He spoke at a session titled "The Global Biodiversity Assessment: The Importance of Biodiversity and Ecosystem Functioning." Mooney asked if the loss of biodiversity affected the function of ecosystems. A new science of biodiversity and ecosystem functioning, Mooney suggested, would integrate ecosystem and population studies. We believe that to be of any value such studies must take place from the vantage point of the landscape. We agree with Mooney that:

There are already compelling reasons to protect the diversity of life, but policymakers need more solid data to help them make hard choices about the consequences of decisions about issues such as land use.

When an ecosystem is fragmented and disturbed, Mooney added, the flow of materials between sections of the landscape can be both disturbed and enhanced. So far, attempts to reconstruct ecosystems have not been totally successful. Ecologists and environmental engineers have tried to fully restore a marsh and they cannot do it, Mooney said. "They cannot get it back to the way it was originally." We share his desire for more complete answers and sense the urgency in his plea.

Hansson and Angelstam (1991) emphasized that landscape ecology could act as a theoretical basis for conservation, an idea suggested by Harris (1984). The need for an appreciation for functional landscape ecology was championed by Merriam (1988). The same physical space has been occupied by a sequence of changing species ensembles over time. Biotic life has not been dictated by physical features or characteristics solely. We must therefore add the time dimension to Rowe's (1988) description of landscape ecology as the study of terrain ecosystems.

Organisms of the present are inextricably tied to organisms of the past by the process of biological evolution taking place over time. The fabric of physical space is woven together by the biological processes of movement and dispersal and physical processes dependent on the biota for their propagation, such as fire (Harris et al. 1996). Species interact in communities and move across space. The ecology of landscapes we embrace extends Merriam's (1988) functional landscape ecology whereby organisms, as propagators of ecological functions, interact with, alter, and maintain their local environments. The functional pillars for our top-down approach are multilevel species evolution integrating through time space that is horizontally connected by dispersal and movement of organisms, processes such as fire that depend on the biota, processes such as hurricanes that act stochastically, but regularly, and processes that occur irregularly and unpredictably such as volcanic eruptions and meteoric impacts. From our perspective any human activity that disrupts functional connectivity across landscapes or handicaps natural evolution must be considered a threat to biological diversity.

A Note on Scale

Bissonette's (1997) discussion on scale is excellent and we will not repeat it here. We argue that while landscape-level effects are scale dependent, ecological processes are invariant and therefore scale independent. For example, individual grass clumps are impediments to terrestrial beetles, while bison merely walk on these same clumps. We suggest that though the quantitative distance traveled by beetles and bison differs according to their size, the ecological process of movement occurred in both species. A landscape effect gave rise to different quantitative outcomes according to the size of the organisms and their mode of travel. Thinking of processes as verbs (to move, to disperse, to evolve) we see that the outcomes of processes are scale dependent, but the processes themselves are scale invariant. After all, the process of addition is scale invariant, but the outcome or effect of performing this process depends on the numbers involved and so is scale dependent.

Another important landscape process is fire. Fire is an example of an ecological process that does not scale across landscapes: the same fire kills trees, damages a patch of trees, and yet is simultaneously necessary for perpetuating the forest. The process of burning is scale independent, but the results depend on the magnitude or scale of the fire.

2

An Epistemology of Landscape Ecology

Jim Sanderson and Larry D. Harris

CONTENTS

We all think we know much about the natural world. We hear, feel, see, taste, and smell parts of the world around us nearly continuously. Furthermore, we all accept that with knowledge of the present and the past coupled with induction, we could generate new knowledge about the future. Obviously, since the sun has risen in the east at least since the time of recorded history, the sun will most likely rise in the east tomorrow. But is this enough? How do we know when something is true? Must we prove something is true before we accept it as fact? What is the source of knowledge and can we develop a theory of human knowledge? The philosophical examination of human knowledge is referred to as epistemology (*Encyclopedia Americana* 1994). As a division of philosophy, epistemology is the study of the origin of knowledge and attempts to develop a theory of the nature of knowledge (*American Heritage Dictionary* 1985).

As scientists, ecologists attempt to explain scientific facts and observations concerning ecology and to create theories that suggest new and novel facts or observations. That is, based on a set of observations, ecologists attempt to suggest further observations that have not yet been made. A theory is created based on a finite set of observations and then used to predict new observations that can be made that have not been made before. If the theory is sound, these new observations will be correctly predicted. A scientific theory in ecology can never be proved by observations, but can be disproved by only one

1-56670-368-9/00/$0.00+$.50

observation. Indeed, the most dangerous threat to a new theory is newer data. Thus, the science of ecology advances not by proving new theories, but by disproving already existing theories. To provide a sound basis for our investigations we must develop landscape ecology as an epistemological paradigm. We follow Scheiner et al. (1993), but change their pattern of linkages between ecological entities to better suit our needs. If the objects, processes, and theories can be woven together into a tight fabric, a solid theoretical basis for making predictions will result.

Definitions

The four basic elements of the epistemology are *entities, processes, properties,* and *theories* and these must be defined (Scheiner et al. 1993). Entities are objects or groups of objects. An individual, a population of individuals, a community, and an ecosystem are examples of entities. Entities can have sub-entities. For instance, a community can consist of numbers of individuals of different species. Abiotic resources can be an entity. Entities are linked by processes that are interactions between objects. The uptake of nutrients between the soil and green plants is an example of a process. Herbivory, predation, mutualism, and movement are examples of processes.

A property is a characteristic of an entity. Properties of entities are familiar to all of us. For instance, one molecule of oxygen when bonded with two molecules of hydrogen is a solid at or below $0°C$, a liquid between 0 and $100°C$, and a gas above $100°C$. The process of heating the entity causes the state to change from a solid to a liquid and then to a gas. Here we must know the context of the molecule to appreciate its form. Individuals have properties such as the ability to fly, populations migrate occasionally, communities such as wiregrass and longleaf pine are thought to form an association that promotes the process of fire, and ecosystem properties are defined by the process of energy flow and the cycling of materials. Landscapes too have properties. Recalling our definition that a landscape is two or more ecosystems, landscapes can support species that require forage in a grassland ecosystem and cover in a forest ecosystem.

We accept Bissonette's (1997, pp. 17–23) definitions of collective and emergent properties. Briefly, if a property can be explained *fully* by mechanisms examined at the next lower level of complexity, then the property is collective, otherwise it is emergent. Epistemological questions such as whether complex systems exhibit emergent properties can be resolved as a matter of definition (Dobzhansky et al. 1977). We cannot tell whether water is a solid, liquid, or gas simply by examining a molecule of oxygen and two molecules of hydrogen, for instance. Landscape processes such as fragmentation are obvious to many of us, but we consider the effect fragmentation has on organisms to differ from the top-down forces organisms have on the landscape.

A theory is a model of how a process acts on an entity to determine the properties of the entity. An example of a relatively new theory is that the sun, as opposed to the earth, is at the center of the solar system. Until 400 years ago, humans believed that the earth was the center and that the sun and other planets revolved around the earth. The new model based on the process of gravity seems to better explain the orbital properties of certain entities called planets, especially those now believed to be beyond the orbit of the earth. This new theory cannot be proved, but is now widely accepted by everyone — as was the previous theory. Theories can be inductively or deductively deduced.

Deductive reasoning is the process of inferring from the general to the specific. Lindeman (1942), after his fieldwork on small Cedar Bog Lake near the University of Minnesota, deduced that the lake was an entity called an ecosystem that could be described as a network of processes or interactions within groups of subentities called organisms linked by the process of feeding and contained within the ecosystem. Lindeman described the interaction between the biotic and abiotic entities of the lake envisioned by Forbes (1887) and others. The lake ecosystem through its subentities took in energy from the sun and nutrients and recycled them into insects and food for terrestrial animals.

Lindeman's theory asserted that nature was organized into ecological systems that were recognizable objects such as lakes that have an origin and development leading to a steady state or dynamic equilibrium. These systems, Lindeman asserted, have a structure defined as a network of feeding relationships among their species populations that can be simplified by grouping the populations into food chains or trophic levels. An ecosystem process, beyond development through time, was that energy received from the sun went into heat and work to process chemical elements. The structure and function of the ecosystem was expressed mathematically as a series of equations describing the interactions between system components. Lindeman claimed, for instance, that the ratio of transfer up the food pyramid varied from 10 to 22.3% depending on the trophic level. Lindeman, 7 years following Tansley's definition of an ecosystem, provided an example and stated a theory that enabled testing of hypotheses and hence defined a program that occupied ecologists for the next 40 years (Golley 1993). Such was the power of Lindeman's Trophic–Dynamic Theory.

Inductive reasoning allows generalization of a class after reasoning about a particular member of the class. For instance, if we observed that all swans in a particular lake were white we might induce that all swans everywhere were white and propose the White Swan Theory. Further observations allow testing hypotheses derived from the theory. Note that only one counterexample is needed to show the theory is incorrect. Thus, induction is the process of creating hypotheses from observations (Briggs and Peat 1984).

Ecological Theories

Though we and others (Diamond and Gilpin 1984) do not subscribe to the theory that nature is necessarily hierarchically ordered, a simple hierarchy will suffice for the purpose of describing where entities fit into our schema of ecological systems (Figure 2.1). First, there exists an entity called a landscape

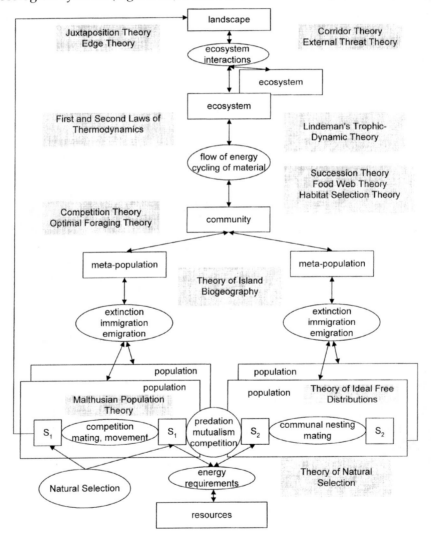

FIGURE 2.1
Many ecological theories have been formulated. Here we emphasize a dualistic viewpoint where interactions occur in both top-down and bottom-up directions.

that consists of two or more subentities referred to as ecosystems. A landscape is therefore made up of an abiotic component and a biotic component. The existence of a landscape depends on solar energy and living organisms interacting on landforms.

An ecosystem is an entity that consists of an abiotic and biotic community that are linked together by the flow of energy through the subentities and the cycling of resources such as water and nutrients. The subentities are communities of organisms, air, soil, water, and other physical resources. The process of energy flow through an ecosystem is constrained by Lindeman's Trophic–Dynamic Theory that can be considered a special case of Newton's Second Law of Thermodynamics. The cycling of material is constrained by the conservation of energy and matter described by Newton's First Law of Thermodynamics. Together, these theories provide powerful organizing principles for the study of ecosystems.

Communities are entities that consist of subentities called populations or metapopulations of organisms. Processes such as predator–prey interactions have been described simply by the Lotka–Volterra equations or more complex formulas, and these descriptions suffice as a theory to generate new facts. The Optimal Foraging Theory (MacArthur and Pianka 1966) describes how organisms select prey items, for instance.

Metapopulations are entities made up of geographically separated subentities called populations. The Theory of Island Biogeography (MacArthur and Wilson 1967) quantifies the process by which metapopulations interact. Recolonization is a metapopulation phenomenon. The processes of migration, extinction, and emigration act on populations that are subentities of metapopulations.

An entity called a population consists of similar subentities referred to as individuals. The Malthusian Principle (Malthus 1798) also acts at the level of the population. Populations have collective properties such as sex ratios, average numbers of offspring, age and size structure, and gene frequencies.

Andrewartha and Birch (1984, p. 185) created a Theory of the Distribution and Abundance of Animals. Before stating the theory formally they carefully described what they meant by the terms model, hypothesis, experiment, and explanation, in effect giving their theory an epistemological foundation. Earlier they purposed the Theory of the Environment that attempted to describe the distribution and abundance of species (Andrewartha and Birch 1954). Indeed, the study of ecology in their book was the study of the distribution and abundance of particular species of animals. Andrewartha and Birch (1954) considered food, weather, other animals, and "a place in which to live" as the four cornerstones of their theory. Later, Andrewartha and Birch (1984) described "the Web" where both direct and indirect factors influencing the distribution and abundance of animals were considered.

When populations of individuals of the same species are linked by dispersal we refer to the ensemble as a metapopulation (Levins 1969, 1970). Metapopulation Theory is now well developed (Hanski and Gilpin 1996;

Gilpin and Hanski 1991; Harrison 1991; Stenseth 1980; Brown and Kodric-Brown 1977).

Pulliam (1988) was one of the first to address the effects of habitat-specific demographic rates on population growth and regulation. In heterogeneous landscapes, source populations and sink populations coexisted and persisted simultaneously. A sink population was a population whose death rate exceeded the survival rate, while a source population was a population that grew. Sink populations were supported by immigration from source populations and could be larger than the source population. The confounding implications for reserve design were also presented. Suppose that 90% of the favorable habitat was occupied by a sink population and consequently what appeared to be poorer habitat was actually occupied by the source population. Saving the largest portion of the total habitat while eliminating the remaining apparently poorer 10% would eventually doom the species to extirpation. Furthermore, Pulliam's Source–Sink Population Theory suggested that diversity and relative abundance of organisms in any habitat might depend as much on the regional diversity of habitats as on the diversity of resources locally available. That is, organisms might nest in an area where nest sites were available, but feed in another where resources were more plentiful. The implications were that studies of organisms often needed to be done within a landscape context.

The Theory of Natural Selection states that individuals are the basic entities upon which selection operates. Individuals have life history strategies and traits. Solar resources, planetary forces, landscapes, ecosystems, communities, metapopulations, and populations all act upon and select against individuals. Individuals act upon landscapes, ecosystems, communities, and so forth, and their resources. For instance, early biotic life on the earth changed the primeval atmosphere of gases deadly to life to one that supports the biodiversity we see today. Thus, a simple hierarchical structure must give way to a fully interconnected, perhaps dualistic structure where all entities eventually select against individuals and all individuals influence all other components of the environment including landforms, landscapes, and even physical processes such as fire and solar insolation (Figure 2.1). After all, although life cannot affect the sun directly, life can and does influence the atmosphere of the earth. For instance, all plants and animals respire, and this affects the chemical composition of the atmosphere.

What is missing from our more dualistic organization of life and resources are the theories that describe the processes acting on ecosystems that give rise to and maintain landscapes, an essential entity in the epistemology of landscape ecology. Without theories that describe the processes that affect entities and their interactions, entities become virtual entities, that is, they cease to exist in any form except our imaginations.

Rather than simply state the theories of landscape ecology we prefer to develop them more thoroughly and then formulate a theory as a result of our investigations. But what are landscape theories? To establish a framework and point of view, consider the following example. Lindeman's

Trophic–Dynamic Theory addresses trophic interactions within a bog lake ecosystem. The lake is necessarily an open system connected to the landscape by the natural atmospheric processes and the mammals, birds, insects, and other organisms that use the lake. Suppose for the moment that we could change the context of the lake from an open system to a partially closed system. That is, suppose we would build a wall around the lake and put a net over the lake to prevent living organisms from either entering or leaving the system. Sunlight, precipitation, and respiration would be unaffected. Lindeman's theory dealt with the contents of the lake. A landscape theory must deal with the context of the lake. Is context important? An hypothesis might be that if the bog lake is isolated from the present context, the contents of the lake would change, perhaps dramatically. The entire trophic system would measurably be altered. If our hypothesis turns out to be true, then we can formulate a Contextual Theory that might state more specifically why context is indeed important to maintaining the trophic structure of the lake ecosystem. Two examples serve to illustrate the value of theory. Each example is a study that required years to complete. By taking a top-down approach we ask how each study fits into the larger picture of ecology. What theory can we induce from isolated examples?

Examples of Landscape Ecological Studies

Leach and Givnish (1996) documented species losses from fragmented prairie remnants in Wisconsin. Prairie once covered an estimated 800,000 hectares in Wisconsin. Today native prairie occurs over a much reduced areal extent in isolated fragments within a fire-suppressed landscape. Recall that fire is a landscape process that depends on the biota for existence. The Theory of Island Biogeography could be invoked to explain the loss of species in isolated prairie fragments. The theory says that as fragments become smaller, species losses will increase. Even the casual observer will admit that prairie fragments in a human-dominated agricultural landscape differ from oceanic islands, however. Several processes might therefore contribute to local extinctions.

Fragmentation increases extinction rates by reducing colonization by similar or different species, for instance. Population sizes probably decrease as fragments become smaller, thus raising the probability that an infrequent event might cause local extirpation of species. Keystone species might be lost and hence alter the ecological balance of the fragment. Edge effects increase and penetrate further into fragments than large contiguous areas. Leach and Givnish suggested, however, that the pattern of plant species loss was consistent with the effects of wildfire suppression. In other words, prairie fragmentation resulted in the loss of large contiguous areas able to propagate fire and by human fire suppression. With the loss of fire certain plant species displaced other less competitive species. Leach and Givnish predicted that with the loss of fire local loss of short species would occur coupled with an increase of tall or woody species. The authors showed that species losses were consistent with predictions derived from the Theory of Island Biogeography. They also cited the loss of forb understory and short- and medium-height grasses in unburned, unmowed, or ungrazed fragments in Nebraska. At Konza Prairie, Kansas overall plant diversity decreased when intervals between fires increased, and increased when annual burning was instituted.

Biodiversity decreased when the prairie was fragmented because of the loss of fire, a key process. What happened when forest-stream processes were disrupted? The disarticulation of a stream from forest litter inputs was studied by Wallace et al. (1997). No physical fragmentation occurred, but the vital process of litter input to the stream was disrupted. By now the consequences of this disruption should be predictable — changes in species abundance and composition and species loss were to be expected.

Inputs of detritus from nearby forests into the headwaters of many eastern North American streams exceeded within-stream primary production (Webster et al. 1995). Wallace et al. excluded detritus input along a 180-m stretch of stream using an overhead canopy and a lateral fence for 3 years. The authors observed major changes in abundance, biomass, and production of invertebrates in the stream. Of 29 major taxa, 17 showed reductions in abundance or biomass, or both. With the loss of primary consumers, predatory species declined in abundance and biomass.

Wallace et al. were led to conclude that processes such as logging, land-use change, fire, grazing, and channelization that reduced terrestrial litter input to streams lead to reduced stream productivity. Though the propagation of the loss of detrital input through the food chain to predators was very much a bottom-up effect, the disconnection of terrestrial inputs to the stream was a connectivity issue because some processes depended on connectedness.

These two examples illustrate that decoupling landscape components whether the same (as in prairie fragmentation) or different (forest and stream disarticulation) had profound consequences for biodiversity because landscapes processes that depended on physical connectivity were disrupted or severed entirely. Rather than rely on the crutch of the Theory of Island Biogeography in a terrestrial setting we prefer to introduce Juxtaposition Theory to enable prediction and hypothesis testing of the consequences of fragmentation or disarticulation. Our theory, stated formally later, suggests that the loss of a key process or processes results in the loss of biodiversity. We must first discover what the key process or processes are that are diminished or lost entirely, understand the agents that perpetuate the process or processes, and then predict how biodiversity will be altered as a result. Often, cascading effects occur that obscure end results; nevertheless we should not be discouraged by the complexity of nature.

Top-down and bottom-up controls have been debated by ecologists for decades. In 1992, the journal *Ecology* had a special feature on the relative contributions of top-down and bottom-up forces in population and community ecology. Hunter and Price (1992) argued that a synthesis of the roles of top-down and bottom-up forces in terrestrial systems required a model that encompassed heterogeneity among species within a trophic level and differences in species interactions in a changing environment. Their model was more suitable for ecological time rather than evolutionary time and hence they argued that a "bottom-up" perspective provided a better first approximation of real pattern in nature. Hunter and Price stated:

> Cataloguing the outcome of single-factor studies is not synthesis. Ecologists tend to champion their favorite ecological factor (indeed some have made careers doing so), but collecting examples of where natural enemies, climate conditions, or primary producers dominate particular systems, and weighing their relative importance by the number of manuscripts in support of each, tells us little about the way the world works.

Though we live in ecological time, we must plan for evolutionary time. Landscape ecological studies must be framed in a 4-dimensional space–time context. What humans arbitrarily define as species in the current time snapshot will change with the passage of time. The difference between evolving into a new "species" and becoming extinct in ecological time is not subtle. The evolution of bison in North America is well documented (Guthrie 1970; McDonald 1981). There are many intermediate forms between ancient and modern species. How have these now extinct "species" been labeled? The species distinction is a discrete name placed with hindsight on an example (or average example!) produced by a continuous process, that of speciation. Moreover, this continuous process produces different results across space. Only cladograms have branching points. Understanding the full complement of genetic variation within species across space lies within the purview of landscape ecology. Landscape ecological studies can extend well beyond the ecological time constraints imposed by the patch-matrix-corridor paradigm. However, we do not intend to use evolution or climate change as a crutch to provide support for landscape ecological studies. The passenger pigeon, whose numbers were estimated to be one-quarter of all North American birds, did not evolve out of existence — humans hunted the bird into extinction.

A Note on Chaos Theory

Potts (1997) suggested that landscapes were becoming increasingly fragmented as early as the Miocene, 12 MYA. The evolution of humans, Potts argued, was a result of the ability of early hominids to adapt to fragmented, increasingly heterogeneous environments. As the Great Rift Valley of Africa was forming, landscape fragmentation accelerated, pushing the precursors of humans to evolve. If Potts is correct, this suggests that humans evolved in the continuously fragmenting landscape of East Africa and that humans were coadapted to the process of fragmentation. Certainly no one would argue the premise that humans are doing well in the most fragmented environment the earth has known. Indeed, fragmentation and heterogeneity are so ingrained in humans that we are apparently compelled to fragment. The problem is not one of fragmentation then (that is one way speciation occurs), but the rate of fragmentation that is of concern.

Modern species other than humans evolved in the presence of fragmentation as well. Mammalian Miocene forms would be easily recognized, and some of their lineages are with us today as different species having evolved in fragmenting environments. So long as fragmentation occurred naturally, species were able to evolve. However, as fragmentation rates increased, biological evolution could not act fast enough. Extinction rates have increased dramatically as a direct result of anthropocentric activities, two of which are habitat fragmentation and destruction. The step-function increase in extinction rates is reminiscent of results from Chaos Theory. That is, things seem to be going along fine until a "critical value" is reached at which time a large change occurs. An example is water at different temperatures. When $0°C$ is reached from above, water makes a physical state change from a liquid to a solid. From 0 to $100°C$ water is a liquid. Another state change occurs at $100°C$. Between critical values, nothing unusual happens. Perhaps the same is true of fragmentation. As humans accelerated fragmentation, a critical value of fragmentation was surpassed whereby evolution could no longer produce forms modified to survive in rapidly fragmenting environments.

Huffaker Revisited

Ecology is a vertically integrated science. Often, reviewing past literature from the vantage point of accumulated knowledge is productive because new conclusions can be drawn. Consider the classical experiment performed by Huffaker (1958) that has obvious implications for reserve design (Andrewartha and Birch 1984, p. 119). Although Huffaker's now classic experiments were designed to test predator–prey interactions, he was the first to test the response of a predator-prey community in a variety of reserve configurations. Huffaker (1958) started with a simple environment and was led to create a series of increasingly complex environments to prolong the length of species coexistence so that their interactions could be studied. Huffaker's experimental evidence suggested that complex heterogeneous landscapes with corridors connecting isolated landscape fragments were essential to prolonging the coexistence of both predator and prey.

Huffaker's experimental methodology and design were created to enable the establishment of an ecosystem where a predator–prey community would continue to exist so that their population dynamics could be studied in detail (Huffaker 1958). From a modern perspective, Huffaker studied reserve design configuration and connectivity of various combinations to prolong the coexistence of a predator and prey community. Previous experiments by Gause (1934) and Gause et al. (1936) led to the conclusion that prey species required a "privileged sanctuary" or periodic reintroductions to continue to exist. Based on experimental evidence, Gause concluded that predator–prey systems would self-annihilate, that predators overexploit their prey, and that in nature immigration of prey was necessary to repopulate local environments where extirpation occurred. Recall that Gause employed two species of paramecium and that his experiments took place in a test tube.

Nicholson (1933, 1954) and Winsor (1934) criticized Gause's broad conclusions because of the simplicity of the microcosm Gause employed. Continuing the reductionist's approach to experiments, DeBach and Smith (1941) attempted to isolate other variables such as "searching capacity" for testing. Because of the force of predation in the "limited universes employed" by others, Huffaker (1958) decided subsequent experiments must be made "progressively more complex in nature and the areas larger."

Huffaker (1958) chose as the prey species the six-spotted mite, *Eotetranychus sexmaculatus*, and the predatory mite *Typhlodromus occidentalis* as his predator–prey community. Since Huffaker was motivated to study mite infestations in agricultural orange groves, he chose as the food supply for his prey species oranges wrapped in paper with only a portion of one hemisphere exposed. Throughout his experiments he used various combinations of oranges and rubber balls uniformly spaced in trays. Barriers and corridors were varied to control the complexity of the environment. Oranges were replaced on a predetermined schedule whether or not they were rotted so as to avoid making the prey food supply a limiting variable in most instances. From a conservation perspective, we are interested in understanding the factors involved in each experiment that determined the coexistence time of the community. We also hope to understand better the difference between conserving one large area vs. several interconnected small areas (SLOSS).

For the initial experiment four lint-covered oranges each with one-half of a hemisphere exposed were arranged in a square pattern in a universe of rubber balls on a tray 40 in long and 16 in wide. Lint was found to provide a physical environment favorable to mite propagation and provided a limited amount of surface complexity that added to the searching time of the predatory species. Immigration and emigration from the tray were prohibited. Controls with no predatory mites were used to establish prey density in the absence of predation.

Three universes were used to establish prey densities in the absence of predators. In the first universe, four oranges with one hemisphere exposed were grouped in a square and connected with a wire loop. Mean density reached 4700 mites per orange area. In the second universe, four oranges with one hemisphere exposed were randomly dispersed on the holding tray. The mean density level per orange area was 3500 mites. The difficulty of dispersal between food supplies had a slightly stabilizing affect on the population, however. Usually, one orange was fully exploited before mites left to seek other food sources. In the third universe, 20 smaller areas of food (one-tenth of the orange exposed) were alternately arranged between rubber balls. The mean population density was 3300 mites per orange area. Differences in the three experiments occurred because food supplies were replenished on a regular basis. When oranges were removed, portions of the population were removed with them.

Nine experiments were used to study predator–prey relationships. The first universe was identical with the previous first universe, there being four grouped, interconnected oranges each with one hemisphere exposed. After introduction and establishment of the prey species, two female predatory mites were introduced on day 11, a standard used throughout most of the experiments. Within 10 d the prey were reduced to very low densities so that all predators eventually starved. A gradual increase in the prey population then occurred. Between introduction and extinction of the predators, 35 days had passed. In the second universe eight oranges with one hemisphere each exposed were arranged to be adjacent and interconnected. Though prey dispersal occurred throughout the universe, the predators totally annihilated the prey species and then followed them to extinction. Subsequently, Huffaker found that older female mites with less reproductive vigor were used to stock the original prey population whose population as a result did not grow rapidly initially. The second experiment lasted 25 d. Although there were confounding factors, the large contiguous reserve did not support the community.

The third experiment employed six fully exposed oranges arranged adjacently and interconnected. Again the predatory mites eliminated the prey and hence themselves. Predators survived for 51 d. Experiment four consisted of four randomly dispersed oranges with one hemisphere exposed. The wide dispersal of food presented both predator and prey with obstacles to movement. Prey species reached a density of 4056 per orange area, but this level could not be maintained on a single orange and subsequent intense predation and exhaustion of the food supply led to abrupt prey population crashes. Predators survived the crash, but eventually died out at which time the prey species populations grew. Predators lasted 52 d. In experiment five, eight randomly dispersed oranges with one hemisphere exposed were employed. Here again all predators starved in 35 d and prey populations subsequently grew.

Experiment six with two replications consisted of 20 alternating oranges each with one-tenth of the area exposed. While in both replicates the predators starved in about 35 d, only in one replicate was the prey species reduced to extinction. Experiment seven had all 40 feeding stations with 1/20th of the orange area exposed. In this experiment the tray was divided into three equal areas mostly, but not entirely, separated by Vaseline™ barriers as an impediment, but not an exclusion to movement. The prey species was introduced on all oranges while only one orange was inoculated with the predatory species. An initial sharp decline in predators occurred after they consumed the prey in one-third of the tray where they were originally introduced. A sharp increase in predators occurred when they reached the second third of the tray where prey populations were at maximum density. Eventually the predatory species starved in 32 d and prey population began a gradual increase.

Experiment eight used 120 oranges each with 1/20th of the orange area exposed and no rubber balls. The Vaseline barriers created a more complex universe than previously employed. The universe was started by placing ten female prey mites on each of two oranges in a single tray. Later, two female predatory mites were added to two of the previously seeded oranges. The action of the predatory mites was delayed; however, movement of the prey from the occupied tray to the next never occurred and both predator and prey went extinct. The experiment ran for 55 d, but predators survived only 37. Experiment nine was an attempt to remedy the lack of prey dispersal that occurred in the previous experiment. Though the configuration was the same as that previously used, small wooden posts were placed in upright positions in each of the major sections of the universe to give the prey species increased opportunity to disperse over Vaseline barriers by utilizing their ability to drop silken threads and be carried by air currents to new locations. Unlike the previous experiment, however, a single female prey mite was introduced on every feeding station and 27 predatory mites were introduced on 27 oranges "these being representative of all major sections of the universe." An electric fan was also used to create mild air movement. Although the predatory mites had superior dispersal abilities, they did not use air movement. However, the division between the trays was a more effective barrier than Vaseline within trays.

Huffaker's Conclusions

In the large and more complex environment predators had a more difficult time locating concentrations of prey at all positions simultaneously, and thus the total population of prey cycled. The predator population responded quickly to any increase in prey densities. Pockets of prey were able to escape predation because the initial distribution of predator introductions and shortages of food kept their numbers from increasing substantially. Many oranges were underutilized, however. Eventually the predators went extinct and the prey began an increase. In this experiment predators lasted 205 d, far longer than any other.

In the final experiment, neither predator nor prey were food limited in the initial phase of population increases. In the three population crashes that followed, predators survived only the first two at single arenas and in extremely low numbers. Previous experimental replicates demonstrated that predator survival was a matter partly decided by chance occurrences. Environmental complexity enabled many locally high prey densities to exist and so the probability of at least one predator population surviving was higher.

To design an environment with enhanced long-term survival potential, the prey must survive the exploitation by the predator. Environmental complexity must be such that the prey are able to colonize areas where they were previously extirpated by predators. In experiments where food resources were concentrated the predators drove the prey to extinction more often than not and then went extinct themselves. Simply increasing the area of odd resources did not remedy this situation. Huffaker concluded that it was "logical to assume that increased complexity is a more important element of the prerequisites than increased area or quantity of food for the prey." Such complexity provides more opportunities for prey refuge against overexploitation by the predator and also reduces the predator's effective reproductive capacity by lowering their density.

Modern Interpretation

Huffaker concluded that his most complex environment was not complicated enough to sustain both prey and their predators. For a perpetuating system sufficient areas must be incorporated to assure several or many arenas of "last survivors" of predators. Huffaker envisaged many "undergirding, larger, nearly self-sustaining subuniverses or ecosystems, each one as adequate or more adequate" than his last experiment. An environment must be large enough to permit the continued existence of both predator and prey, yet not so large that the population in its arenas can proceed asynchronously, "due to too limited interchange of the biotic participants." He also noted that if the area is sufficiently small and simple, biological parameters "which have been present during the long evolutionary origins of the relations involving the participants" would be absent and the interactions between the species would be dysfunctional.

Dispersal and movement were key processes that Huffaker maintained were important to maintaining a functioning system. Unlike Gause (1934), Huffaker realized that emigration from outside the system was not necessary to sustain species interactions. Barriers acted to delay contact of prey by predators, but once the prey were discovered their population was able to sustain a larger predator population. This might have been offset if resources were more limiting. Barriers created the "partial asynchrony in geographic position" and promoted earlier population recovery, and this feature was critical to maintaining the predator.

Huffaker also concluded that the balance or stability observed in nature was characteristic of the total environment in which evolution of a predator–prey relationship occurred, and this community would tend to be stable over the long term. He noted that in large, simple monocultures the inoculation of a single predatory species often solved the most severe pest problems. We too are interested in maintaining both predators and prey over the long term. The results obtained by Huffaker lead to the conclusion that to support a predator–prey community, many small, irregularly interconnected reserves function better than a single large reserve or a series of regularly interconnected reserves where predators drive their prey to such low densities that the predators themselves cannot survive. These results suggest that environmental heterogeneity and connectivity are essential in maintaining predator–prey populations.

EXERCISES

2.1 Alaskan brown bears are well known to eat salmon and also feed in berry patches. Predict the response of the surrounding stream vegetation to the decline or loss of salmon to Alaskan streams. Identify the key process or processes that are likely to be lost. How can you test your hypothesis?

2.2 Nonnative Lake trout are replacing native Cutthroat trout in Yellowstone Lake, Yellowstone National Park. Lake trout spawn in deeper parts of the lake while the native trout spawn in shallow feeder streams. Predict the response of Yellowstone grizzly bears and use this prediction to discuss changes in the composition of the surrounding vegetation. Identify the key processes being lost.

2.3 Wolves are being reintroduced to Minnesota. Predict their indirect impact on forest mid-canopy and ground birds.

3

The Presence of the Past

Jim Sanderson and Larry D. Harris

CONTENTS

In his book *Zoogeography*, P.J. Darlington (1957) attempted to explain the history and distribution of vertebrates across the world using widely accepted principles and practices of his time. Plate tectonics was not widely accepted, so explaining the distribution of freshwater frogs and fishes presented some interesting difficulties. Darlington appreciated that each continent changed shape and form, and was disconnected and reconnected to close neighboring continents through time. Though intercontinental seas grew and drained and surface details changed, Darlington believed that the continents remained anchored in their present positions and their climates remained nearly constant at least for the last 200 million years.

Vertebrates, Darlington explained, dispersed across barriers by rafting and by wind. Powers (1911) described a 100 ft^2 land raft with trees 30 ft. high that was seen along the Atlantic Coast of North America in 1892 that was known to have traveled at least 1000 miles. Over the course of millions of years thousands of such rafts could have transported the ancestors of elephants from Africa to North America for instance. Once land was touched most vertebrates could make their way across a continent in tens of thousands of years if each successive generation spread by a few meters.

The power of wind as a dispersing agent was also described by Darlington. Strong winds, he suggested, could transport small organisms, say a mouse weighing 100 g, invertebrates, or mollusks. Darlington's arguments remain interesting and provocative, though we now have much more plausible explanations for the distribution of organisms. Understanding the past and

1-56670-368-9/00/$0.00+$.50
© 2000 by CRC Press LLC

appreciating its presence on a landscape is often a critical component of landscape ecological studies.

We must also accept that humans influence every ecosystem and landscape on earth and humans are here to stay at least for the foreseeable future. Landscapes, however, are ephemeral and the time frame of their change is shrinking as the human population increases and more of the natural landscape is converted into human food production. Therefore landscape studies will most often include the impacts of humans if only indirectly. Even studies of snow leopards in Mongolia where the human population density is less than 1 person per 5 km^2 take into account the impact of humans and their life-support systems. Indeed, though the entire human population of earth could be contained entirely in Alaska with each family of four having 1/5 hectare, the full impact of collective humanity would still be global in extent. Evidence of the production of human food, extraction of natural resources, industrial creation of products, communication infrastructure such as satellites, and air pollution is found in even the most remote, inhospitable, uninhabitable, and historically unoccupied areas of the earth. Our theory of landscape ecology, of necessity, must therefore include human processes that act upon the biota. Indeed, our preoccupation with studies in landscape ecology is no less than to understand and then restore natural processes operating in ecological and evolutionary time in the presence of human activities that operate typically over much shorter time periods. For example, Hastings and Turner (1980) documented the roles played by humans and climate in altering the arid landscapes of the Southwestern U.S. They used comparative photography to illustrate the profound changes in stream courses and vegetation, some of which occurred in a few tens of years.

Theodosius Dobzhansky wrote "Nothing in biology makes sense except in the light of evolution" (Dobzhansky 1973). To deduce a theory of landscape ecology we must synthesize theories that describe here-and-now processes controlling ecosystems and adopt an evolutionary approach to ecology that permits the inclusion of past events. For instance, Lindeman's Trophic–Dynamic Theory describes the processes structuring Cedar Bog Lake. The theory neither describes how the lake came into existence, nor purports to tell us what the lake will look like in 5000 years. The study of most ecosystems is very much a here-and-now activity. Indeed, bottom-up processes such as nutrient flow act relatively quickly, while soil development takes place over long periods of time. Ecosystem studies generally take place over time spans of months or years and only rarely over decades. Therefore, while the Trophic–Dynamic Theory allows deduction about the processes structuring an ecosystem we cannot accurately induce a theory of landscapes based upon such a theory. That is, while we might be able to generalize the Trophic–Dyanmic Theory to many ecosystems, we are unwilling (or unable) to induce a theory of landscape ecology from a single theory.

Our plan here is to adopt a top-down approach to develop a theory of landscape ecology. To achieve our purpose we begin with a description of external processes operating on the earth itself. Certainly some of these forces are beyond human intervention. Plate tectonics, for instance, are acting continuously, but no less powerfully in time. Volcanoes are geological as well as ecological events that have often instant and profound impacts on ecosystems, landscapes, and biomes, for example. Biogeography textbooks (Brown and Gibson 1983; Darlington 1957) thoroughly cover the origin and spread of life throughout evolutionary time. For a summary see Gould et al. (1993). Because most of what we see of the natural landscape has been highly altered, we will begin with a review of some of the processes that have been acting on landscapes through time.

Cosmology

No discussion of life on earth can be complete without mentioning the sometimes dramatic effect extraterrestrial bodies have had. Life on earth is inextricably tied to random events occurring in the rest of the solar system and indeed beyond (Norton 1994). That large meteorites have impacted earth in the past is now widely accepted. Fortunately, most objects that enter the atmosphere burn up well before impact. That all earth impactors had negative consequences is debatable. Vershuur (1996) suggested that a dozen or so massive comets striking the earth could have provided enough water to fill the basins of ancient earth and seeded the water with organic molecules.

A typical planet, including earth, is formed from the remnants of supernovae of massive stars. The envelopes of aging supergiant stars are production factories producing organic molecules, many that are carbon based. Over vast time scales these molecules could have spread across galaxies, some transported by comets.

Our moon exerts a gravitational force that acts as a lock on the axis of the earth allowing excursions around 23 1/2°. Without this soft lock on the axis the earth would wobble dramatically in cycles lasting millions or hundreds of millions of years. All life on earth lives under the random influence of comets and asteroids (Kelley and Dachille 1953). Certainly the K-T boundary collision 65 MYA is the best known well-documented event. We now know the Permian extinctions of 250 MYA occurred rapidly, wiping away 85% of all marine species and 70% of all vertebrate genera on land. Insects suffered their only known mass extinction. Life on earth was very nearly extinguished completely (Bowring et al. 1998).

By 1995, 145 impact craters had been discovered. About 30 were buried, covered over by time-dependent geological processes. One 50-MY-old impact crater 45 km in diameter was found in the sea off Nova Scotia. Another is suspected to be in the Barents Sea. One need only casually observe the surface of the moon to see what a thin veneer of an atmosphere can do. The K-T meteor that struck the coast of what is now the Yucatan Peninsula,

Mexico was estimated to be 10 km in diameter. This is pitifully small compared to the earth yet if one end of the meteorite was stuck into the ocean the other end would protrude 6 km and be among the highest peaks on earth.

Patterns on the Martian surface show that water once flowed there. Perhaps the Martian atmosphere was blown into space by a large impactor. Martian gravity is weaker than that of the earth and might not have been able to pull the atmosphere back. An object 200 m in diameter striking the earth would do severe regional damage and perhaps planetary-wide damage. The impactors that struck Jupiter in 1994 were estimated to be each less than 1 km in size and 1/50th the density of water. Each of the 21 fragments was not a single well-defined object, but more like an aerosol or powder. Each fragment produced an atmospheric fireball of gas heated to perhaps 10,000°K, twice as hot as the surface of the sun. The fireball spread out of the Jovian atmosphere at 17 km/sec. The vastly expanded plume then cooled and rained back into the upper atmosphere at 5 km/sec, heating volumes of the atmosphere many times larger than the earth to 2000°K. The implications of a similar event on earth can only be imagined.

In 2126, the comet Swift–Tuttle will pass through the orbital plane of the earth. Described as the single most dangerous object known to humanity, comet Swift–Tuttle will be around for perhaps 20,000 years. The comet was first documented in 1862 and rediscovered again in 1992. Before that it was observed in 1737. In 1993, the orbital parameters of Swift–Tuttle were changed when material was ejected. Crude estimates suggest that the comet will pass within 23,000,000 km of earth in 129 years. This translates to a margin of safety of about 14 d assuming no other ejections take place over the next century and a quarter. Swift–Tuttle is about 24 km in diameter and so far larger than the K-T impactor and is traveling about 61 km/sec. Estimates suggest that the impact energy would equal some 5 billion megatons compared to the estimated 100 million megatons of the K-T impactor, an impact energy beyond human comprehension.

Biotic Responses to Abiotic Change

Just as organisms have evolved throughout time, so have landscapes. Plate tectonic movements changed the position of the land masses and the orbital parameters of the earth–sun system changed slowly in time, causing ice ages (Imbrie and Imbrie 1979). Life responded to and evolved as a result of these changes. Plant and animal communities have responded to changing climatic patterns in a Gleasonian fashion. Davis (1969) and Graham (1986) showed that vegetation and animal communities, respectively, were ephemeral. Just as communities evolved through time, so did ecosystems. Delcourt et al. (1983) showed that a 60- to 100-km-wide tundra existed in front of the great Laurentian Ice Sheet 18,000 years ago. This tundra extended, not continuously, down the Appalachian crest to the Great Smoky Mountains. Warm-temperate forests of oak, hickory, and southern pine grew near the Gulf Coast and lower Atlantic coastal plain.

As the climate warmed and the ice sheet retreated, the tundra migrated northward, oak savanna developed in Florida, and open prairie occurred in the Great Plains of today. The largest body of water in North America was Lake Bonneville in part of Utah, Nevada, and Idaho. During the middle Wisconsin the lake was 330 m deep and nearly 50,000 km^2. Today, we know part of this region as the Bonneville salt flats. Lake Chad in northern Africa covered over 300,000 km^2, but is now only 16,000 km^2. The disappearance of the pluvial lakes in the Great Basin of the western U.S. had profound implications for plants and animals highly dependent on bodies of fresh water as well as the regional climate where the moderating effect of a large water body was lost.

The Milankovitch Theory states that variations in three orbital parameters of the earth are responsible for changes in climate. As with any theory, testable predictions were enabled and also verified. The Milankovitch Theory not only predicted the number and location of previous ice ages over the last 650,000 years, but also predicted the future course of the climate of the earth. For instance, the theory states that, driven by changes in the orbit of the earth, a cooling trend will begin approximately 2000 years from now. In 3000 years what remains of the oak forests of Europe will disappear and the longest interglacial period will come to an end. 23,000 years in the future earth will be engulfed in another age of ice. How was such a predictive theory developed?

Louis Agassiz was not the first to grasp the significance of large boulders strewn in various, widespread agricultural fields across Europe, but in 1837 he was one of the earliest well-known scientists to suggest the large blocks of rock had been transported to their present position by glaciers (Imbrie and Imbrie 1979). His arguments were not universally accepted. No less a scientist than Alexander von Humbolt urged Agassiz to return to the study of fossil fishes.

By the mid-1860s the idea that the earth had indeed experienced an ice age was widely appreciated. While geologists continued to gather evidence on a past glacialization, various theories were created to explain causes of an ice age. For instance, a decrease in the output energy of the sun might have given rise to an ice age. An astronomical theory of the ice age was first proposed in 1842 by the French mathematician Adhémar. Croll (1886) presented a more complete astronomical theory that could be tested by comparing predictions with the geologic record. By the late turn of the century, however, the astronomical theory was largely discredited by the rapidly accumulating geologic record. Not until 1924 with the publication of Milankovitch's work was the full power of the astronomical theory revealed.

Milankovitch used changes in the eccentricity, precession, and tilt of the earth to create a powerful mathematical theory he used to predict past and future glacial and interglacial periods. His theory told geologists not only in what rock layers to look, but also where to look. Nevertheless, doubts about the theory lingered into the 1970s. We now know that three million years ago ice sheets occurred for the first time in the northern hemisphere. This ice sheet and the occurrence of extensive land areas in the northern high latitudes were sensitive to astronomical variations in the orbit of the earth. Following the formation of the ice sheet a series of glacial and interglacial periods began. Their duration and extent were predicted by astronomical theory of the ice ages and are now supported by the geologic sea floor record. Cycles of 100,000 years (eccentricity), 41,000 years (axial tilt), and 23,000 and 19,000 years (variations in precession) are found in the climate record of the earth and clearly support the Milankovitch Theory.

Though the installation of thousands of weather stations surrounding the globe were important to making weather predictions, weather patterns differ from climatic patterns. Weather occurs over far shorter time periods and within a regional climate. The value of the Milankovitch Theory was that future temperatures, for instance, could be predicted for each weather station around the world whereas all the assembled weather data ever collected could not be used to predict the future climate or create a meaningful theory of the climate of the earth. Similarly, the results from thousands of studies of ecosystems, while important to interpreting local ecological phenomena, have not been used to induce a theory of landscape ecology. However, these numerous studies can be used to support or refute such a theory once one is proposed.

Evidence for global repeated alternations between glacial and interglacial periods exists back to the mid to late Pliocene when, at least once, temperatures in Antarctica were 15 to 20°C warmer than today (Cronin and Dowsett 1991; Harwood 1985; Mercer 1987; Webb and Harwood 1991). These temperatures allowed birch *(Nothofagus)* to exist in the Transantarctic Mountains. The geographic range and distribution of vegetation varied with climate change. Hooghiemstra and Sarimento (1991) examined pollen records between 4 and 2.5 million years ago in the high plains surrounding Bogotá, Colombia. Fully 26 alternations between Páramo vegetation (warm assemblages) and Andean forest vegetation (cold) occurred.

During the late Pliocene the Isthmus of Panama closed, permitting the great American interchange (Stehli and Webb 1985). Biotic organisms from the North American continent migrated to South America, and likewise some South American organisms migrated northward. For instance, *Alnus* evolved in North America and migrated southward, reaching Colombia by 2.7 MYA, and today is found in southern Argentina. Similarly, *Quercus* arrived in Colombia one MYA and has a current southern limit in Ecuador. However, Northern Hemisphere temperate forests species were not able to colonize South America so that the present *Nothofagus* forests of the Southern Hemisphere are of very ancient lineage (Markgraf et al. 1996).

Profound changes in the faunal and floral components of the biota occurred in North America as well. For instance, between 12,000 and 9,000 YA nearly 40 large mammals went extinct in North America. Several hypotheses have been advanced to account for these extinctions (Pielou 1991). Graham et al. (1996) analyzed 2945 fossil late Pleistocene and Holocene mammal occurrences in the U.S. Species range shifts occurred that corresponded with changes in the climate. Two models have been suggested to describe species range changes. The Clementsian model suggests that entire communities responded to change similarly. That is, species ranges shifted successionally north and south as the climate alternated. The Boreal community, for instance, would have moved northward with the retreating ice.

The Gleasonian model assumes that species responded in ecological time to environmental change according to their individual requirements and not as communities. Graham et al. (1996) found that species range shifts occurred with different rates, at different times, and often in divergent directions, supporting the Gleasonian theory. In other words, the mammal communities were indeed ephermal and continuously changing in response to habitat reorganization, biological interactions, and stochastic events often in unpredictable ways.

The implications of climate change on reserve design were thus profound. The floral and faunal communities we see around us today will not be here 25,000 years from now during the next ice age. Clearly, migratory paths for species that require them will be necessary to enable communities to disassociate and reassemble elsewhere naturally. However, the case for corridors to assist the biota to migrate in response to climate change is a weak foundation for claiming that landscape ecology merits serious study in its own right. After all, climate change did not cause Florida panthers, manatees, Florida beach mice, Cape Sable seaside sparrows, snail kites, and Florida scrub jays to be classified as Federally Endangered Species. Furthermore, climate change has not been linked to modern (post-1600) extinctions.

The natural environment of the earth is continually changing, and climate change is a relatively slow natural process that can be accelerated by humans. Naturally occurring episodic events such as fire, volcanoes, mud slides, and the like are processes that affect the biota as well. The boreal forests of the northern hemisphere are fire maintained, for instance. The Yellowstone fires of 1988 affected much of Yellowstone National Park and were presumed to be large. We suggest that because Yellowstone is a relatively small area compared to the greater Yellowstone landscape, the impact of the fire merely appeared large. That is, because fires were suppressed in Yellowstone, fuel loads increased, and because Yellowstone is too small, the fire appeared quite large. In all likelihood, more widespread fires occurred in the natural landscape more frequently than in modern times and the forest community was adapted to such fires.

Denslow (1985) divided effects of episodic events on ecosystems into three components: changes in environmental heterogeneity, changes in temporal heterogeneity, and changes in the relative abundance of species. Certainly, different physical conditions and biotic components contained within a landscape determine the effects of and effects on episodic events. Pickett and White (1985) suggested "How a particular disturbance event may alter a system or affect the distribution of resources, and the distribution and coexistence of species in an area, will be greatly influenced by the nature of the landscape." The topographic heterogeneity, environmental gradients, and various barriers within a landscape effect and help define episodic events. Indeed, predictions of episodic events must also appreciate that the particular context may enhance or constrain the outcome of an event. Pickett and White (1985) recognized that the following factors were important to understanding episodic events: (a) system structure, (b) resource base, (c) life and natural history parameters, and (d) landscape composition and configuration. We add to this list landscape context. While any theory of landscape ecology embraces episodic events, these events alone cannot be used to establish a paradigm for the study of landscape ecology. Our view necessitates a shift in focus from the contents within an area or region to the processes acting in the contextual setting of the area. Moreover, dynamic cross-boundary processes are seen as being critically important.

Human Evolution

The origin of modern humans has been traced back by independently analyzing mitochondrial DNA and, separately, polymorphisms in the male Y chromosome (Gibbons 1997). Both studies led to the same conclusion: a man and a woman living in Africa 100,000 to 200,000 years ago were found to be ancestral to all living humans. Living men whose Y chromosomes most resembled that ancestor were found to be Ethiopian, Sudanese, and Khoisan people (Hottentots and Bushmen) living in southern Africa. The presence of the past is carried in our nuclear DNA.

Africa has been drying and cooling for at least 1 million years. Species turnover in bovids, rodents, and hominids has been correlated to this long-term climate trend. Indeed, the origin of bipedal hominids and later toolmaking and increased brain size have all been linked to the cooling, aridification, and expansion of savannas in Africa. Potts (1997) suggested that an increase in the range of climate variation occurred. Pleistocene records from Africa, Europe, and Asia indicated repeated shifts in regional hydrology and vegetation. Expressed over intervals of 10,000 to 100,000 years the effects on African landscapes were profound and likely had strong influence on water, food resources, biotic competition, and thus natural selection. Potts proposed the term "variability selection" to indicate that selection was not unidirectional. Variability selection would favor pliable gene combinations.

As global climate variability increased between 3 and 5 MYA, a form of bipedality emerged that could accommodate terrestrial and arboreal settings. Increase in relative brain size, technological innovation, and geographic diversity in behavior were all apparent in the late Pleistocene. The emergence of these features corresponded with the greatest recorded oscillations of the late Cenozoic climate. Some archaeological sites in Africa showed environmental change with corresponding changes in fauna, but also showed continuous occupation by hominids. Apparently these hominids were able to cope with landscape changes. Five previously dominant herbivores in southern Kenya that specialized on low-quality forage became extinct between 800,000 and 400,000 years ago. They were survived by related taxa better able to switch diets, i.e., those lineages with more flexible genes. Later, hominids such as Neanderthals that specialized in geographic locations and cold climates (Neanderthals had short distal limbs, for instance) became extinct whereas those with greater mobility and behavioral diversity such as **Homo erectus** and **H. sapiens** persisted. The hominids of 100,000 to 200,000 years ago were well adapted to climate change. Perhaps this explains in part why humans are able to survive and thrive anywhere on earth. Because humans evolved during a time of high climate variability and increasing fragmentation, humans can be considered fragmentation specialists. Indeed, humans continue to prosper through increasing fragmentation. Perhaps we are compelled through our genes to fragment our environment.

Natural Variation in Space

Through time and across space organisms varied nearly continuously. If, for instance, all horses that ever existed throughout time and space could be brought together for study, the differentiation to species level would be nearly impossible. For a given time, horses varied across space. Moreover, at a single place horses varied in time. Today we have ring species, that is, species that vary as gulls do, for instance, across the northern hemisphere. Although geographic extremes look quite different, intermediate forms may not be different enough to be reproductively separate. Ring species demonstrate that, just as individuals vary, populations vary, and this interspecies variation leads, by extrapolation, to questions about just what a species really is (Ridley 1993). Placing organisms into bins called species is a human construct that helps to organize our thinking. Species are therefore analogous to the natural numbers. We count things by using integers. Between any two integers there are a countably infinite number of rational numbers (fractions) and also an uncountable number of irrational numbers. By our analogy then, through space and time, species have varied. The idea that organisms come in discrete bins is therefore a gross simplification.

Within a species a cline is a gradient of nearly continuous variation in a genotype or phenotypic character. Houses sparrows, for example, show a cline in size with respect to geographic location in North America (Gould and Johnson 1972). A cline can result when evolution favors certain genotypes in different environments such as occurs across landscapes. Mayr (1963) refers to differences among spatially separate populations of a species as *geographic variation.* He considered geographic variation "a universal phenomenon in the animal kingdom." Humans show great geographic variation. The widespread incidence of geographic variation in plants (so-called ecotopes) has been appreciated since the 1950s. Some snails show extreme localization of phenotype. Mayr (1963) stated that "any character, be it external or internal, morphological or physiological, may vary geographically." The Pampas cat is considered to be a single species of cat through its range in South America. Rose-Peria (1994) argues, however, that variations in skull and dental features demonstrate that the Pampas cat should be considered three separate species and two subspecies.

We should not dwell exclusively on variation in space at a given time. Chronological variation (chronoclines) in populations has also been documented. But what do chronoclines and geographic clines (choroclines) have to do with landscape ecology? Is most variation irrelevant and void of biological significance? Is the close correlation between the environment and geographic variability merely coincidence? Are the ecogeographic rules of Allen, Bergmann, and Gloger relevant and, if they are, how does this variation affect the ecology of the species in question? Moreover, does the ecology of the species vary across space? Behavioral and ecological variation across space is relevant to landscape ecology and the study of such differences amounts to taking advantage of natural experiments. We can, in a sense, fix the "species" and vary its environment to elucidate its ecological significance across a landscape or region. Each population is the result of continual selection processes. The genotype of each local population has been selected for a locally adapted phenotype. Perhaps the study of different locally adapted populations can help uncover differences in the ecology of a species and hence lead to a better understanding of the ecology of landscapes.

EXERCISES

3.1 Waterfowl populations were in serious decline at the turn of the century. A series of stepping-stone reserves was established as migrating waterfowl feeding stations. Using Canada geese as an example, investigate the effectiveness of this conservation strategy. What is happening with Canada geese populations today?

3.2 Bison are being restored to parts of the Dakotas and especially on Native American lands. Predict the possible impacts on the fauna and flora of the region. Recall that bison are keystone species and that they are migratory. We might think of bison as creators of landscape connections.

4

Landforms and Landscapes

Jim Sanderson and Larry D. Harris

CONTENTS

Understanding landforms and geomorphological processes is necessary to understanding the ecology of landscapes (Swanson et al. 1988). Rainshadows caused by the interaction of wind and mountains are but one obvious example of why landforms are important to the biota. Landforms effect the flow of energy and material that support life. Landform gradients can deliver water to wildlife or act to constrain their migratory paths. Appreciating the influence of landforms and climate on soil, vegetation, and wildlife is a familiar bottom-up path to gaining an appreciation for landscape ecology.

Landforms and Geological Processes

The relief sphere or outer boundary of the crust of the earth is the boundary between the atmosphere and the solid crust or lithosphere. About 71% of the relief sphere is covered by oceans. Geomorphology is the study of the genesis and history of landforms, or more precisely, of the *subaerial relief sphere*, presently 26% of the surface of the earth exposed directly to the atmosphere (not covered by ice or sea) where higher plants and animals evolved and where humans live (Büdel 1982). The shape of landforms is actively molded by plate tectonic volcanic and exogenous processes last derived ultimately from solar energy that is converted to heat energy mainly at the surface of the earth. The appearance of landforms varies with climate.

The relief sphere of the earth is covered by an atmosphere some 800 km thick. All processes of radiation, wind movement, pressure waves, and other changes in the atmosphere, and all surface and deep-water currents, tidal movements, swells, convection, and internal waves in the ocean are ephemeral processes that leave little or no lasting evidence in the atmosphere or ocean. Only in the crust of the earth and sea sediments are past events sometimes recorded. Before Rutimeyer (1874) demonstrated that valleys were not cracks in the crust, but instead were hollowed by water, the relief sphere was thought to be formed by purely endogenic processes such as crustal movements (Lyell 1830 to 1833). We now know that two often related endogenic processes have direct influence on landform relief, and these are associated with volcanic events (cones, craters, and calderas) or plate tectonic movements. In both cases exogenic processes work to change the external appearances of landforms.

Processes that affect landforms, so-called relief-forming mechanisms, derive their energy from short-wave radiation impinging upon and heating the relief sphere. While radiation penetrates water, on land the conversion is intense and confined to a relatively thin layer. Whereas water mixes heat rapidly and thus distributes energy over a greater extent, landforms are heated directly and the heat is transported downward, but no deeper than 10 to 12 m below the surface where annual temperature fluctuations cease. All processes associated with weather and climate are therefore governed by the relief sphere. Climatic effects set off far-reaching energy conversions that create relief-forming mechanisms that ultimately shape landforms.

Relief-forming mechanisms are classified as either stable elements such as insolation, evaporation, and precipitation or mobile elements such as erosion, transportation, and deposition. The relief sphere is struck by insolation and warmed, causing evaporation and hence precipitation, the influence of which depends ultimately on the form (solid or liquid) taken. Evaporation is a proximate effect, giving rise to deserts and, together with wind, causing other areas to receive moisture. Moist air is transported by the atmospheric currents and precipitation is deposited according to well-understood principles. Precipitation can penetrate the earth and cause mechanical and chemical changes. Water is the main agent causing bedrock decomposition that aids in soil formation. In the humid tropics water causes chemical weathering and in the high latitudes water in the form of permafrost causes mechanical change, breaking down rock *in situ*.

The mobile relief-forming mechanisms represent the final stage in energy conversions of landforms and transform the endogenous form into the exogenic landforms we see today (Figure 4.1). Areal and slope erosion are the most obvious examples of mobile mechanisms, and the great valleys that cover many areas have resulted from erosion. Indeed, most relief development is involved with creating concavities (Büdel 1982). Material is worn

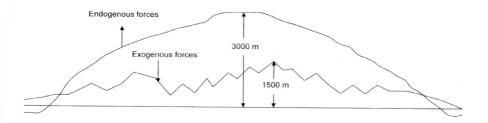

FIGURE 4.1

Endogenous tectonic forces have created a dome-shaped mountain that has risen over tens of millions of years. The mountain's internal structure is highly folded due to unequal uplifting forces. The exogenic actual shape is jagged and worn down, transformed into valleys whose material wears faster than do neighboring peaks. Landforms we see today are being actively shaped by exogenic mechanisms, such as erosion, transportation, and deposition.

down and then transported by wind or water elsewhere. Rivers deliver suspended material to the sea, and while some material is dropped quickly, the finest material such as clay can be carried far out to sea and deposition takes place in the deep ocean. The erosional power of wind is also well known. In the Botku Mountains of Chad, 700 km northeast of Lake Chad, wind has stripped away the soil and is abrasively wearing away the exposed bedrock. Wind is also a depositional process. Great Sand Dunes National Monument in southern Colorado exists because wind carries fine sand particles, sorting them for many kilometers, and depositing the nearly uniform grains into great piles before rising over the San Juan Mountains.

With few exceptions, the landforms we see today are the products of an ancient past. The Sahara Desert once resided at the magnetic South Pole and was covered by ice. India changed hemispheres. The Himalayas continue to rise upward and be eroded downward. Land now free from kilometers of thick ice rises and will sink during the next glacial advance. Tidewater glaciers advance and retreat. Oxbow lakes are formed and reformed as water from the Andes finds the sea. These processes acting on landforms at differing time scales create the stage upon which the thin veneer of biotic life exists. The theater that contains the stage is also dynamic, but on much longer time scales.

Tricart (1965) created a table classifying geomorphological features. For our purposes Table 4.1 links the time required to effect change on an area of the surface of the earth by well-understood processes. The importance of Table 4.1 from an ecological viewpoint is that many landscapes have changed dramatically even over the last ten years principally by the acts of humans and not by geological or lithographic processes. In other words, in creating a reserve or protected area we should not dwell on how the landform beneath the biota will change. Instead, anthropocentric effects are occurring over much faster time spans and should be of central concern.

TABLE 4.1
Geomorphogic processes operate on time spans that change landform features. Generally, the larger the area affected the longer terrestrial processes require to effect change. (From Tricart 1965.)

Area (km²)	Example	Process	Time span (years)
10^7	Ocean basin	Plate tectonics	10^9
10^6	Scandinavian shield	Crustal movements	10^8
10^4–10^2	Mountain massifs	Smaller tectonic movements	10^7
10	Valleys	Lithography	10^6
10^2	Moraines, ridges	Lithography, wind	10^4
10^6	Badland gullies	Microlithography	10^2

Though the composition of the atmosphere has been changed by the biota, the circulation patterns are governed by planetary forces such as the Coriolis Force, and the resulting Hadley, Ferrel, and Polar cells. The flow of atmospheric currents and the three-dimensional form of the relief sphere determine the location of the deserts, rainshadows, and wet areas even in the absence of life (Figure 4.2). Global ocean currents are also unaffected by the biota. The Coriolis Force, surface winds, ice sheets, and the location of the continents determine ocean currents. The ocean transports energy in the form of heat to the high latitudes, from the low equatorial latitudes moderating the global climate. Though surface winds may change direction, the long-term climate memory of the earth lays within the ocean currents. Atmospheric winds and ocean currents continuously recycle air and water. Moisture-laden air arising from evaporation over the oceans is transported by surface winds over land where precipitation occurs. Ocean water recycles CO_2 and other gases in the air. Indeed, the ocean bottom is the largest storage of carbon in the form of limestone ($CaCO_3$) on the planet. Living organisms with shells formed the limestone sequestering the carbon.

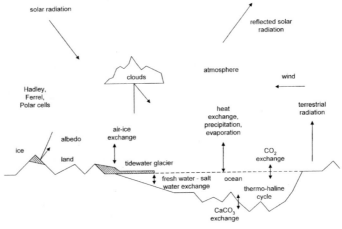

FIGURE 4.2
The climate system of the earth would still exist even in the absence of life.

Relief-Forming Mechanisms and Their Effect on the Biota

The stable and mobile relief-forming mechanisms exert their impact on the biota over continuous spatial and temporal time scales. The physical forces of plate tectonics, Milankovitch cycles, and volcanoes occur in the absence of life and appear insensitive to the biota in terms of their intensity, recurrence interval, and rate of change. Their effects on the biota, however, are profound. For instance, we can predict with near certainty that further deglaciation and sea level encroachment onto the Florida peninsula will cause further extinction pressure on coastal species in the future.

A second example involves the intensity, recurrence interval, and behavior of precipitation and wind patterns known as hurricanes. Under existing climatic conditions, approximately 100 tropical waves develop in the Atlantic Ocean off the West Coast of Africa annually. Perhaps one third of these develop into tropical depressions and, on average, six of these will develop into hurricanes. At least one of these will develop into a substantially damaging hurricane (i.e., class 3) somewhere within the 12 M km^2 Caribbean region in any given year. The probability that such a storm will occur in the more limited 3.5 M km^2 area encompassed by the West Indies in any given year is considerably less. The probability that an equal-intensity storm will hit the 9000-km^2 island of Puerto Rico in any given year is yet less, and the probability that a large hurricane will directly affect the dozens of endangered species occurring in the Caribbean National Forest are very much less. Though the chance that a major episodic event such as a hurricane will impact any given area is directly related to the size of area considered, the overall significance of the event varies inversely to the size of the area (Figure 4.3). Such observations have profound implications for the design of nature reserves (Harris et al. 1996).

FIGURE 4.3
The probability of a hurricane striking the Gulf Coast each year is 1.0. For each successively smaller stretch of coast the probability of a hurricane striking decreases. Conversely, as the probability of a strike decreases, the damage done if a strike occurs increases. Therefore, the probability of a hurricane striking a beach reserve decreases with the areal extent of the reserve, however, the amount of damage is inversely proportional to the size of the reserve. A small reserve or protected area is unlikely to be struck by a hurricane, but if struck the reserve will, in all likelihood, be destroyed.

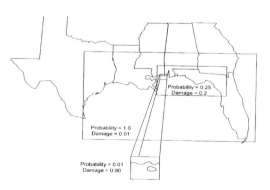

Biotic Processes

Life does not simply exist on landforms, however. Throughout the evolutionary history of life, biotic organisms have acted to transform the atmosphere and landforms. Mechanical weathering (mobile decomposition) breaks down rock without chemical changes. Chemical decomposition such as hydration can completely change the parent rock. Weathering and the oxygen present in water lead to oxidation of iron and aluminum compounds, giving rise to the brown and red stains observed in many soils. Indeed, soil formation cannot take place without the biochemical influences of the biosphere. Soil development is an important process performed by the relief-forming mechanisms of most climatic zones and biotic life. At least one physical process, fire, requires terrestrial biota to occur and propagate (Bond and van Wilgen 1996). Prior to the existence of land-based life and an oxygen-rich atmosphere, fire did not exist. Fire, therefore, is an emergent process of the biota.

The profound influence biotic life has had on the formation of the atmosphere is well known. Only recently has the influence of ocean algae on the atmosphere been elucidated (Hansson 1997). Algae contain methionine and convert that compound into dimethylsulfoniopropionate (DMSP) as protection against freezing and high salinity. DMSP is then converted to dimethylsulfide, a sulfur gas, that is a significant product involved in the global sulfur cycle. By seawater–air exchange, dissolved dimethylsulfide becomes atmospheric dimethylsulfide and oxidizes into sulfuric acid. Minute sulfuric acid particles enter the atmosphere and aid in cloud formation and also act in the upper atmosphere to reflect heat energy into space, cooling the earth. Ocean algae not only moderate, but in part create the climate of the earth.

The majority of natural ecological systems begin with the process of fixation of solar energy by green plants called autotrophic organisms, or autotrophs, using photosynthesis that is the formation of carbohydrates from water and carbon dioxide. The actual gain in tissue and energy that is accumulated in plants is termed net primary production. Losses due to the process of photorespiration depend primarily on temperature. Nutrient return to the soil begins with the process of litter decomposition. Nutrient release by decomposition of organic wastes occurs by the process of humification, an oxidation process in which complex organic molecules are broken down into simpler organic acids that eventually are mineralized into simple inorganic forms that can be sequestered by plants. Aside from the magnitude of transfer in an ecological system, important factors are the turnover period and the turnover rate, both of which are processes already described elsewhere (Schultz 1995). Bormann and Likens (1979) detailed other ecological processes, especially those performed by forests.

The above biotic processes are referred to as ecological system processes or simply ecosystem processes. These processes can and do take place on areas as small as the head of a pin. Because most ecosystem studies are concerned with the fluxes of energy and materials and because the defined boundaries of these studies are sometimes the atmosphere and the soil or, when spatially large, a watershed or even a northern hardwood forest, we adopt the definition of Bormann and Likens (1979) that an *ecosystem* is bound together by a similarity of vegetation structure, species composition, and development processes. An ecosystem is fairly uniform throughout the entire range as Bormann and Likens suggest. An ecotone is the boundary between two ecosystems.

We therefore define a *landscape process* as one that occurs in a spatial domain, is primarily propagated across the land surface by the biota, and not only serves to structure, but is also influenced by the spatial structure of ecosystems, and whose strength of impact on spatial pattern is dependent to a greater or lesser degree upon landscape structure.

Landscape Processes

Animal movement across two or more ecosystems is a landscape process involving a member of the biota moving across ecosystems transporting nutrients, and thus performing bottom-up functions and influencing plant communities, thus performing a top-down function. Fire also has the potential to be a landscape process. Human activities are increasingly landscape, as well as regional, biome, and continental processes. Consistent with the definition of a landscape process, a *landscape* thus consists of two or more ecosystems and necessarily contains an ecotone. Inasmuch as ecology is the study of the harmony of living organisms in their environment, landscape ecology is the study of the ecology of landscapes and necessarily involves the juxtaposition of two or more ecosystems whose ecology acts upon and is in turn influenced by mobile organisms across space and through time. Landscape ecology embraces heterogeneity of vegetation communities and seeks, among other pursuits, to explain how organisms act to maintain heterogeneous environments over space and time. Humans perform processes that affect and are affected by landscapes and so are another organism whose processes require study within landscapes.

Naveh (1991) remarked that the greatest challenge for landscape ecology was to cope with higher levels of organized complexity and their emergent qualities, transcending those of populations, communities, and ecosystems. We must also appreciate the differences between landscape effects and landscape ecology. Why, for instance, is biodiversity so important to preserve (Holling et al. 1995)?

Tansley (1935) in his brilliantly constructed paper first praised Clements and Clements' disciple Phillips, and then proceeded to demolish the Clementsian theory of vegetation leaf by leaf. We will not dwell on these arguments, but instead discuss other relevant points made by the author. Tansley argued that human prejudices forced ecologists to consider organisms as the most important parts of ecological systems while inorganic "factors" received less attention. Organisms and inorganic factors together constituted an ecosystem. Ecosystems formed one category of the multitudinous physical systems of the universe, that ranged from the universe as a whole down to the atom. The fundamental concept appropriate to the biome (the whole complex of organisms) considered in its entirety together with the environment and inorganic factors was the ecosystem. According to Tansley, organisms and inorganic factors were components of an ecosystem.

In a carefully worded sentence, Tansley referred to anthropogenic ecosystems that "differ from those developed independently of man." The essential formative processes of the vegetation were the same, however the factors initiating them were directed. Humans could indeed influence processes and achieve different vegetative outcomes. He argued that a collection of ecological concepts must allow the inclusion of all forms of vegetational expression and activity. Ecologists should not separate the study of so-called natural communities from anthropogenic communities, for the processes maintaining both were identical. There was no fundamental difference between grazing by bison on plains grasses and grazing by cattle, he believed. The effect was the same, though he agreed that by killing carnivores to protect cattle humans acted to artificially change the herbivore community and hence perhaps the composition of the grassland. He also asserted that humans might alter the position of equilibrium by feeding animals not only on the pasture, but also partly away from it, so that the animal dung represented food for grasslands brought from outside, and the floristic composition of the grassland might therefore be altered.

Troll (1971) provided a history of terminology in common usage. He wrote that climate, rock, and water were the basis of plant life. At the same time he realized that plant cover affected water budgets, soil genesis, and the differentiation of microclimates. Communities of plants, animals, and microorganisms, on the other hand, were controlled by environmental factors. Schroter later extended the study of ecology to "synecology," the study of a complete living community in its relationship with the environment. As early as 1931, Berg, cited in Troll (1971), wrote "The Landschaft is at once a community of a higher order consisting of communities of organisms (biocenoses), including plants (phytocenoses), animals (zoocenoses) and occasionally men, together with the complex of inorganic phenomena as for example the form of the relief, the waters and the climatic factors. To the elements of the Landschaft belong also the objects seen by man as derived from both organic and inorganic phenomena of the earth's crust, for example the soil."

Landschaft ecology, Troll suggested, was "the study of the main complex causal relationships between the life communities and their environment in a given section of a Landschaft." Troll derived his view of landscape ecology from aerial photography. Tansley (1939) later defined an ecosystem as "the whole complex of organisms and factors of environment in an ecological unit of any rank." The natural differences that Troll saw in photographs led him to place more emphasis on horizontal considerations. The smallest unit of a landscape, he argued, should be unitary in all edaphic factors and relatively homogeneous in its biological–ecological content. Troll referred to this unit as an ecotope. From Troll's perspective the science of geology considered more regional units of the earth whereas ecology focused on the study of what he called microsites. Landschaft ecology, he argued, brought the two sciences together.

Our definition of a landscape is closely allied with Tansley's (1935) original definition of ecosystem and Troll's (1971) definition arising from the analysis of aerial photography. Tansley's ecosystems are the units of Troll's landscape. That is, we include both organisms and physical factors in our definition of ecosystem and we then extend the concept of an ecosystem by requiring a landscape to be two or more ecosystems separated by an ecotone, and having a large spatial extent. Landscape ecology includes the study of landscape effects that can take place in a 1-m^2 plot. Landscape ecology requires that these spatial extents be studied through time. Our goal is to develop a theoretical basis for landscape ecology that enables predictions to be made. Just as general laws are not deterred by contrary phenomenon (Does every object fall toward the center of the earth or do some bounce once or twice?), counterexamples should not be discouraging when the preponderance of evidence favors one side. However, we are not suggesting that such evidence be dismissed or forgotten.

Zoogeomorphology and the Influence of Wildlife on Ecosystem Dynamics

Butler (1995) summarized the role of free-ranging vertebrates and invertebrates in shaping landforms. The geomorphic effects of animals encompassed their roles in eroding, transporting, and depositing the constituents of landforms, i.e., soil and rock. Examples included the activities of earthworms, termites, ants, arachnids, gastropods, crayfish, fish as burrowers and nest builders in stream soils, amphibians, reptiles, crocodiles, and mammals from aardvarks to wild pigs. Even birds transport mud for nests, or build mound nests, cause erosion, and excavate burrows. Geomorphic effects of mammals have been observed by most of us particularly in arid landscapes where hillsides have been crisscrossed by cattle trails or where beavers have constructed dams and created and maintained lakes (Johnston 1995; Naiman 1988).

The geomorphic activities of animals, however, are just the most obvious evidence of their ability to create long-term effects on landscapes. Processes performed by organisms affect the biota on the landscape and over long time periods. For instance, kangaroo rats live in communal burrows in arid landscapes. Hawkins and Nicoletto (1992) noted that kangaroo rats modify their environment by excavating burrows that lead to mounds that create a relatively moist environment that in turn becomes a focal point for plants and animals. The resulting floral differences from the surrounding grassland can be easily recognized. Kangaroo mounds are long-term features acting to substantially alter the local environment (Mun and Whitford 1990). We will discuss in greater detail the top-down influence of mobile organisms in Chapter 7.

Landscape Conservation

The landscapes of today have been shaped by powerful, ever-present forces now dwarfed in space and time by anthropogenic activities. Whereas landscape evolution was influenced once primarily by natural changes, humans have caused more profound changes to occur in far less time. Mitigating human-caused changes must be achieved with an appreciation for the natural background processes that continue to cause landscape evolution. That is, establishing a national park or reserve, or protecting a biological "hot spot", while minimizing human impacts on a landscape, does little to re-enable natural background processes to function on a dynamic playing field. Furthermore, as Kushlan (1979), Janzen (1986), and others realized, the contextual setting of a park, reserve, or hot spot is far more critical than was once appreciated. Emphasis must be placed on protecting the processes that gave rise to and promoted organic evolution on a dynamic landscape.

EXERCISES
4.1 Does the moon have landforms? What processes are acting on the moon to form landforms? Comment on the Martian landforms.
4.2 Earth Surface Processes and Landforms is the journal of the British Geomorphological Research Group. Compare current issue titles with titles of an early issue. What changes in emphasis have occurred?
4.3 Is a cornfield with a tree windbreak a landscape? Does a cornfield juxtaposed with a wheat field constitute a landscape?
4.4 Does the boreal forest that extends across Siberia, the largest tract of forest on earth, constitute an ecosystem or a landscape? How does an elevational gradient create landscapes? What is the analog of the boreal forest in the Southern Hemisphere?
4.5 What is the minimum size a landscape can have? Should landscapes have open or closed boundaries?

4.6 *Can humans influence regional climates? Consider the Aral Sea in Asia as an example.*

4.7 *Design a reserve to protect sea turtles. Note that 99.99% of their lives are spent in the open sea.*

4.8 *Compare and contrast landscapes and seascapes. Terrestrial reserves should be interconnected. Describe how marine reserves are interconnected. What governing processes are similar?*

4.9 *How has the presence of the other planets in our solar system affected life on earth?*

4.10 *How does the moon affect life on earth? How would life differ without the moon?*

Part II

The Ecology of Landscapes

5

The Ecology in Landscape Ecology

Jim Sanderson and Larry D. Harris

CONTENTS

What is the ecology of landscapes? Why should we study landscape ecology? Landscape ecology is the study of processes and organisms that promote and maintain the natural functioning of more than one ecosystem. When mobile organisms and processes are decoupled from ecosystem processes by fragmentation or loss of connectivity, natural ecosystem processes change, often in catastrophic ways harmful to humans. Species identities, numbers of species, community composition, and physical and chemical processes in ecosystems change. Often, a loss of biodiversity results. The ramifications to conservation of the biological resources of the earth are clear. To maintain naturally functioning ecosystems landscape connectivity must be preserved.

Earlier we wrote that advances in landscape ecology will come from the study of landscape effects and also the effects mobile organisms have on landscapes. Many studies show that the spatial structure of landscapes has important effects. The effects mobile organisms have on landscapes have

1-56670-368-9/00/$0.00+$.50
© 2000 by CRC Press LLC

received relatively less attention. With the emergence of a new appreciation for keystone species (Simberloff 1998) and ecosystem engineers (Jones et al. 1997, 1994; Lawton 1994) and the recognition that multilevel evolution (Wilson 1997) takes place in local communities, we are confident all organisms will be seen as important contributors to the ecology of landscapes.

Landscape Effects

Tilman and Kareiva (1997) presented in detail the importance of spatial structure in population dynamics and species interactions. The collected readings demonstrated repeatedly that space played a pervasive role in determining stability, patterns of diversity, invasion of exotics, coexistence, and pattern generation. Many others in excellent texts on landscape ecology also described the profound influence space has on organisms and processes (Bissonette 1997; Lidicker 1995; Hansson et al. 1995). The fundamental paradigm for the study of landscapes espoused by Forman and Godron (1986) was the patch-matrix-corridor model. Lidicker (1995, p. 15) stated the paradigm for landscape ecology precisely:

> We now have a conceptual framework in which to analyze the influence of habitat patch size and shape, the role of juxtaposition of different kinds of communities, the importance of corridors among patches, the influence of edge effects, and the impact of varying proportions and qualities of different community-types in a landscape.

Hansson (1995) provided an excellent historical summary of research efforts in landscape effects. We believe that spatial effects are not the full story told by landscape ecology. How, for instance, will the quantification and sorting of landscape patterns illuminate functional organization? First, however, we must gain an appreciation for these effects. Here we emphasize that fragmentation is a landscape process that impacts ecosystem processes. In 1985, habitat fragmentation was seen as the most serious threat to biological diversity because fragmentation (1) reduced available habitat and (2) increased insularization (Wilcox and Murphy 1985). We now know that the effects of fragmentation go well beyond those originally envisioned (Harris 1988).

Tropical Forests

Laurance et al. (1997) found that rain forests in Amazonia were not subjected to frequent fires, so soil sequestration of C may not be as important as in other latitudes. However, massive quantities of C were sequested in the biomass of tropical forests. Laurance et al. found that rain forest fragments in central Amazonia experienced dramatic loss of above-ground biomass not offset by recruitment. Forest fragments suffered from enhanced negative edge effects such as increased wind turbulence and microclimate changes. Fully 36% of the above-ground biomass within 100 m of edges was lost in the first 10 to 17 years of fragmentation. This decline in biomass, the authors suggested, could lead to increased CO_2 levels in the atmosphere.

Boreal Forests

Wardle et al. (1997) considered the influence of island area on ecosystem properties. Recall that the Theory of Island Biogeography (MacArthur and Wilson 1967) suggested that occurrence and abundance of species were proportional to island area. If individual species were important in determining ecosystem-level properties then islands with different areas should contain different ecosystem-level attributes. Wardle et al. examined 50 islands of the same age and origin ranging in size from 0.02 to 15 ha found within two lakes in the northern boreal forest zone of Sweden.

The main episodic event on the islands was wildfire caused by lightning strikes. Lightning struck larger islands more frequently than smaller islands. Larger islands had more earlier-successional plant species which dominate in the presence of regular wildfire, whereas smaller islands showed a greater abundance of successional species that occurred in the prolonged absence of fire. Smaller islands had higher concentrations of water-soluble phenolics, reduced microbial biomass, and less microbial activity in humus. Reduced rates of decomposition and mineralization of litter were also found on smaller islands. The inhibition of soil biotic processes on smaller islands probably contributed substantial accumulation of humus found on them. The smallest islands contained ten times more humus than did the largest islands. Wardle et al. concluded that island area was critical in regulating key ecological processes.

With increasing island size there was a distinct trend of an increasing proportion of organic carbon bound in living organisms, especially trees. Wildfire was thus of critical importance in reversing C lockup in boreal forest ecosystems. Deliberate anthropogenic suppression of fires in boreal forests had the potential to lead to retardation of soil biological processes and substantial terrestrial C sequestration. Boreal forests play a globally significant role in the C cycle, and fire suppression likely adds to carbon buildup in the atmosphere.

These results show that ecosystem properties depend on island size and hence are applicable to human-created fragmented forests. As landscapes become disarticulated episodic events such as fire occur less often, leading to changes in vegetation characteristics and community composition that alter ecosystem processes. Wardle et al. showed that global carbon sequestration in soils was enhanced by more frequent fires and so we must be led to conclude that fragmentation leads to increased levels of atmospheric CO_2.

Fragmentation and Bird Communities

The effects of forest fragmentation on forest bird communities was documented by Robinson et al. (1995). Nest predation and parasitism by brown-headed cowbirds (*Molothrus ater*) increased with forest fragmentation in nine midwestern U.S. landscapes that varied from 6 to 95% forest cover. Observed reproductive rates were low enough for some species in the most fragmented landscapes to suggest that their populations were sinks that depended for perpetuation on immigration from reproductive source populations in landscapes with more extensive forest cover.

Many neotropical migrant birds were suffering population declines from causes that may include the loss of breeding, wintering, and migration stopover habitats (Robinson 1993). Habitat fragmentation may allow higher rates of brood parasitism by cowbirds and nest predation (Gates and Gysel 1978; Temple and Cary 1988). Cowbirds lay their eggs in the nests of other "host" species, which then raise the cowbirds at the expense of their own young.

Populations of cowbirds and many nest predators were higher in fragmented landscapes where there was a mixture of feeding habitats and breeding habitats. In landscapes fragmented by agricultural fields, levels of nest predation and brood parasitism were so high that many populations of forest birds in the fragmented landscapes were likely to be population "sinks" in which local reproduction was insufficient to compensate for adult mortality (Pulliam 1988). As landscapes become increasingly fragmented, this reproductive dysfunction could cause regional declines of migrant populations.

Robinson et al. (1995) tested the hypothesis that the reproductive success of nine species of forest birds was related to regional patterns of forest fragmentation in Illinois, Indiana, Minnesota, Missouri, and Wisconsin. Cowbird parasitism was negatively correlated with percent forest cover for all species. Most wood thrush (*Hylocichla mustelina*) nests in landscapes with less than 55% forest cover were parasitized. In some landscapes, there were more cowbird eggs than wood thrush eggs per nest. In contrast, cowbird parasitism levels were so low in the heavily forested landscapes that cowbird parasitism was unlikely to be a significant cause of reproductive failure (May and Robinson 1985).

Levels of nest predation also declined with increasing forest cover for all species. Three ground-nesting warblers, the ovenbird (*Seiurus aurocapillus*), worm-eating warbler (*Helmitheros vermivrus*), and Kentucky warbler (*Oporrnis formosus*) and two species that nest near the ground in shrubs, the hooded warbler (*Wilsonia citrina*) and the indigo bunting (*Passerina cyanea*) all had extremely high (6% or higher) daily predation rates in the most fragmented landscapes. Of the 13 cases of daily predation rates exceeding 7%, 12 were in the 4 most fragmented landscapes. Fragmentation at the landscape scale thus affected the levels of parasitism and predation on most migrant forest species in the midwestern U.S. In more fragmented landscapes the cowbird populations may be more limited by the availability of hosts and may saturate the available breeding habitat, resulting in high levels of parasitism even in the interior of the largest tracts in Illinois. Therefore, landscape-level factors such as percent forest cover determined the magnitude of local factors such as tract size and distance from the forest edges, a result consistent with continental analysis of parasitism levels (Hoover and Brittingham 1993).

Nest predators such as mammals, snakes, and blue jays (*Cyanocitta cristata*) likely have smaller home ranges than cowbirds and may therefore be more affected by local rather than by landscape-level habitat conditions. Small woodlots in agricultural landscapes had high populations of raccoons (*Procyon lotor*). Censuses in both Missouri and Wisconsin showed blue jay and crow (*Corvus brachyrhynchos*) abundances to be higher in fragmented regions. Parasitism levels of wood thrushes, tanagers, and hooded warblers and predation rates on ovenbirds and Kentucky warblers were so high in the most fragmented forests that they were likely population sinks.

Robinson et al. (1995) suggested that a good regional conservation strategy for migrant songbirds in the Midwest was to identify, maintain, and restore the large tracts that were most likely to be population sources of songbirds. Further loss or fragmentation of landscapes could lead to a collapse of regional populations of some forest birds. Increasing fragmentation of landscapes was likely contributing to the widespread population declines of several species of forest birds.

Spatial structure, however, is not the only factor influencing ecosystem processes. Herbivory, biodiversity, atmospheric connectivity, and climate change also influence ecosystem processes.

Atmospheric Connectivity

Physical processes operating over large areas such as El Niño events are often overlooked when studies of metapopulations in fragmented landscapes are undertaken. One reason might be that ecological studies typically begin as bottom-up enterprises with site selection and organism autecological studies, for instance. Regional processes, however, are gaining more attention in ecological studies. Studies performed in the White Mountains of New Hampshire would be less valuable if the effects of acid rain were not considered. Long-term studies at the Hubbard Brook Experimental Forest, a long-term ecological research station, have shown that vegetation stopped growing in

1987 and that the pH of many regional streams remained below normal (Likens et al. 1996). Prior to the 1970 Clean Air Act and the 1990 amendment, acid rain deposition was blamed for reacting with soil calcium and magnesium, causing trees to cease adding biomass. Moreover, even though acid rain caused by fossil fuel consumption became less of a problem, soils and hence forests were not expected to recover rapidly. The number of species lost in lakes due to acidification may be substantial. A testable hypothesis would be that insect, reptile, and bird populations occupying the forests of the northeast that have suffered acid rain deposition have less reproductive success compared to those living in unaffected forests.

Climate Change and Grasshoppers in Australia

Birch (1957) documented the role of weather in determining the distribution and abundance of the grasshopper, *Austroicetes cruciata*, in Australia. Weather, Birch argued, might also be important in determining the qualitative composition of a population, affecting survival and reproductive capabilities. Birch was careful to add that factors other than weather were also important in determining the numbers of many animals. Local weather patterns are, in part, determined by landscape features.

The distribution of the grasshopper in the southwestern portion of Australia where outbreaks of the insect occurred were delineated by rainfall and evaporation contours. That is, north of a particular soil moisture contour the soil was too dry to support grasshoppers. South of another contour, the soil was too wet for grasshoppers. Birch showed that the contraction and expansion of the grasshopper belt in a north–south direction was associated with fluctuations of weather. Previously, Andrewartha and Birch (1954) were able to estimate where outbreaks were likely to occur given moisture data from previous years. Previously, an outbreak had occurred in 1937, but had disappeared by 1938 and for several years thereafter.

Birch suggested that fluctuations in weather and other components of the environment and spatial patchiness determined whether or not outbreaks occurred. Understanding spatial heterogeneity and variable weather conditions over time were crucial to predicting when and where grasshopper outbreaks occurred. By considering the demographic characteristic of the grasshopper and tying their life history stages to the environment, Birch argued that weather did indeed play a critical role in grasshopper outbreaks. He strengthened his case adding that grasshopper populations further north had been exterminated by severe weather, while further south small populations were able to survive during unfavorable times, but outbreaks were not known to occur. Any latitudinal changes in weather would favor one population over another so that the species was protected against extinction, at least over ecological time and in the absence of human alterations to the land.

Although establishing a reserve to protect an insect pest is unlikely, the implications of such a task were illuminated by Birch. Had we knowledge of only a few years worth of distribution data the likely possibility exists that we might have established a reserve outside the prime contours of the most favorable areas. The reserve would have to be big enough to encompass areas where grasshoppers are rarely seen except in exceptional years. Suggesting that saving 99.9% of the area occupied by grasshoppers over a limited time period would most likely fail because the vagaries of weather might cause dramatic consecutive range shifts outside protected areas. We would, in fact, be forced to set aside a large heterogeneous, contiguous area with enough north–south extent to include areas that most often do not support any grasshoppers. Preserving a "hot spot" in the center of prime habitat simply would not work for the grasshopper.

The Ecology in Landscape Ecology

The above examples are what we refer to as landscape effects. Specifically, the arrangement of space affects the collective and emergent properties of organisms (Bissonette 1997) and hence ecosystem properties. Fragmentation is a spatial feature that also affects ecosystem properties as does grazing, climate change, and atmospheric transport. However, we believe this work begs the question: What effect do organisms have on the ecology of this space? Answering this question for mobile organisms is fundamental to closing the circle of research on the ecology of landscapes. Johnston (1995) states so clearly and simply the essence of landscape ecology that we need only repeat her opening paragraph slightly modification by omitting references:

> Studies of animal-patch interactions have generally focused on how animals are affected by patchy habitats rather than how they create them. Scientists who observe animal activities that may alter habitat usually quantify and interpret those behaviors in terms of the life history of the animal, rather than consequences to the landscape, hence it is often assumed that their influence is minimal. However, there is growing evidence of the ability of animals to influence landscape pattern and process.

Bowyer et al. (1997) in their opening paragraph agree:

> An increasing body of evidence suggests these large herbivores play a crucial role in determining the structure and function of the ecosystem they inhabit. Moreover, we contend that the role that moose and other large herbivores play in ecosystem processes has been neglected by many ecologists and that future advances in ecosystem science will require integrating

the behavior and population ecology of large mammals into the existing paradigms of landscape ecology.

We prefer to include all mobile organisms into the paradigm. In addition to Johnston (1995) and Bowyer et al. (1997) contributions to this paradigm have been made (Pastor et al. 1998; Nummi and Pöysä 1997; Johnston et al. 1993; Johnston and Naiman 1990; Naiman 1988; Botkin et al. 1981; Zlotin and Khodashova 1980; Harris and Fowler 1975; Belsky 1995). Landscape ecology includes the study of landscape effects on organisms and the top-down effects of mobile organisms on ecosystem processes and the creation and maintenance of spatial heterogeneity. One need only consider the ability of mobile organisms to disperse seeds, for instance, to realize organisms have often profound impacts on the landscape.

Grazing and Ecosystem Functioning

For two years, McNaughton et al. (1997) studied nitrogen (N) and sodium (Na) recycling by nonmigratory grazing herbivores in Serengeti National Park, Tanzania. Grazers preferred forage rich in minerals that were important to late-stage pregnancy, lactation, and growth of young animals. Two hypotheses described this phenomenon: grazers foraged on vegetation supported by nutrient-enriched soils, or grazing augmented nutrient availability. McNaughton et al. found no evidence of general soil differences in study plots used or avoided by grazers that did not migrate.

Concentrations of Na were found to be universally and substantially higher in soils of animal concentration areas. McNaughton et al. concluded grazing increased Na supply from soils by a factor of 10. The net N mineralization rate in soils supporting dense resident animal populations was over twice that of areas where animals were uncommon. Herbivory by Serengeti grazers accelerated mineralization of N and Na, both important in animal nutrition.

Mammalian herbivores co-evolved with grasslands and their grazing accelerated nutrient cycling. Overgrazing of grasslands commonly associated with the replacement of free-ranging wild herbivores with livestock often causes the replacement of highly palatable forages with plant species of lower nutritional quality and decomposibility. McNaughton et al. concluded naturally occurring terrestrial grazers modified ecosystem processes in ways that alleviated nutritional deficiencies. This study also showed that accelerated nutrient cycling was an important property of habitats critical to large-mammal conservation.

Pastor et al. (1998) studied moose foraging and plant communities on Isle Royale, Michigan. Moose have been studied by several research teams in other geographic areas. Because moose are large browsers in comparatively less diverse forests their top-down effects are possible to measure. McNaughton et al. (1997), Naiman et al. (1988), and others found that herbivores alter ecosystem functioning, change species composition in communities, modify nutrient cycling, and hence productivity, thereby changing landscape structure. A high density of moose, $3.7/km^2$, occurred in the study area of Pastor et al. Because moose arrived on Isle Royale at the start of the 20th century their possible effects have taken place quite rapidly.

Other factors such as aspect, topography, and beavers might also have caused patterns in landscapes. Pastor et al. concluded, however, that there were no differences in nitrogen availability or browse consumption due to slope, aspect, underlying bedrock, fire history, or glacial history. Moose avoided spruce and only lightly browsed balsam fir. Thus, these trees were more likely to become large-diameter trees over time. Because their leaves were high in lignin and resin they were slow to decay and release nitrogen in the soil. Where browsing was intense, aspens and other hardwoods were dominated by spruce and balsam fir.

Pastor et al. concluded that the selective foraging by moose caused and maintained both local patches of vegetation and nitrogen cycling rates "as well as the development of higher order patterns across the larger landscape" of the boreal forest. Furthermore, because moose and wolf populations oscillate through time, some properties of the boreal forest many exhibit long-term periodicity. Hansson (1979) suggested that herbivore population cyles inevitably result in heterogeneous resource distributions. As Pastor et al. state:

> Such population cycles and associated spatial patterns may, therefore, be an intrinsic property of an intact, properly functioning ecosystem or landscape. A characteristic of such oscillating systems is not some particular population level, or rate of ecosystem process, or even a particular static pattern in the landscape, but rather a spatial and temporal variance structure of all components of the landscape.

Deer and Songbirds

Species richness and abundance of forest songbirds have been positively correlated with species abundance, composition, and vertical structure of woody and herbaceous vegetation (MacArthur and MacArthur 1961; Karr and Roth 1971; Hopper et al. 1973; DeGraff et al. 1991). Tilghman (1989) and Frelich and Lorimer (1985) documented an inverse relationship between deer density and density of woody vegetation < 1.5 m in height. At deer densities greater than $11/km^2$ species richness and abundance of herbaceous and woody vegetation declined (Behrend et al. 1970; Alverson et al. 1988; Tilghman 1989).

McShea and Rappole (1992) demonstrated a positive correlation between understory vegetation density and songbird species richness and abundance and noted that deer densities were higher in areas with reduced understory vegetation. Casey and Hein (1983) compared differences in bird occurrence and abundance between an area affected by 27 years of ungulate browsing and an adjacent area with lower deer density. Ten species of ground-nesting or intermediate canopy-nesting birds were absent or occurred at lower frequencies in the area with higher ungulate density.

deCalestra (1994) found that mean richness of intermediate canopy-nesting birds declined 27% from the lowest to the highest deer density. Four interior cavity-nesting species (eastern wood pewee, indigo bunting, least flycatcher, yellow-billed cuckoo) were not detected when deer densities exceeded 7.9 deer/km^2 as they were on sites with lower densities of deer. The American robin and eastern phoebe were not detected at deer densities greater than 14.9 deer/km^2 where they had been detected elsewhere where deer densities were less. Cerulean warblers, upper-canopy nesting birds, were not detected at deer densities greater than 14.9/km^2. Richness of ground-nesting birds and other birds in the study area were unaffected, however.

deCalestra (1994) also found that abundance of intermediate canopy-nesting birds declined 37% from the lowest to the highest deer densities whereas ground-nesting and canopy-nesting bird abundance was unchanged. deCalestra concluded white-tailed deer densities greater than 7.9/km^2 reduced intermediate canopy-nesting species richness and abundance by reducing height of woody vegetation in the intermediate canopy less than 7.5 m on thinned and clear-cut sites. Three intermediate canopy-nesting species (Carolina wren, warbling vireo, yellow-breasted chat) and two ground-nesting species (golden-winged warbler, worm-eating warbler) were not present at any of the study sites, but had been previously reported as present. Smith et al. (1993) noted declines in abundance of several intermediate canopy-nesting species in the northeastern U.S., including the eastern wood-pewee, the least flycatcher, and the yellow-breasted chat, species that either disappeared with increasing white-tailed deer density or were absent from deCalestra's study area already. By altering critical nesting habitat for intermediate canopy-nesting species in fragmented forests, where they were already exposed to increased predation and nest parasitism, high deer density further endangered these avian species.

Coherent Landscape Ecology Paradigm

In 1993, Wiens et al. (1993) wrote that a coherent paradigm for landscape ecology had yet to emerge. One approach at integrating landscape dynamics was to study the dynamics of mobile organisms in terrestrial ecosystems. Wiens et al. studied voles, forest grouse, and their predator the pine marten in a heterogeneous landscape. Because these organisms occurred in differing vegetation cover types (boreal forest, grasslands, brush) the consequences of an interaction between martens and voles also affected grouse since the marten hunted in several vegetation types. The species level process of movement was affected by the spatial pattern of the landscape mosaic—again a landscape effect. The Patch Foraging Theory (Stephens and Krebs 1986) considered movement within home ranges; dispersal was the process that took an individual beyond the natal range where so-called patch choices were made. Corridors linked patches in the landscape mosaic and allowed individuals to disperse safely. Since human activities not only alter, but disrupt natural phenomena, these activities must also be considered when studying landscape heterogeneity and the movement of biotic organisms. We will return to this theme in Chapter 7.

Biodiversity

Various studies have shown that the number of species, the number of functional groups, and species composition influence ecosystem processes. Hooper and Vitousek (1997) showed that composition and diversity were significant determinants of ecosystem processes in grasslands. Functional diversity had a greater impact on ecosystem processes than did species diversity. Factors that changed ecosystem composition such as invasion by exotic organisms, disturbance frequency, fragmentation, N deposition, predator decimation, and extinction were likely to strongly affect ecosystem processes.

Protecting Biodiversity "Hot Spots"

In the widely read journal *Science,* Dobson et al. (1997) wrote that "the amount of land that needs to be managed to protect currently endangered and threatened species in the United States is a relatively small proportion of the land mass." Such areas have been termed "hot spots." The authors argued that if endangered species (924 as of August 1995) were highly concentrated, then fewer areas required protection. A computer algorithm was used to locate the maximum number of listed species while minimizing the area of the county (there are 2858 counties in the U.S.) that contained them. For plants, birds, fish, and molluscs, 50% of endangered species were represented within 0.14 to 2.04% of the land area. Southern Nevada, southern California, and parts of Arizona, New Mexico, Florida, and Hawaii contained about 50% of federally listed species. Clear associations existed between the intensity of human economic and agricultural activities and endangered or threatened species. In their closing statement, the authors concluded that "If conservation efforts and funds can be expanded in a few key areas, it should be possible to conserve endangered species with great efficiency."

In the same issue of *Science*, Pulliam and Babbitt (1997) noted that "If we can improve our knowledge of the distribution and co-occurrence of species, then we can provide a sounder scientific basis for ecosystem-based habitat conservation plans (HCPs), cooperative agreements that protect many species under a single plan." Remarkably, in the same issue of *Science*, the demise of the Bay checkerspot butterfly (*Euphydryas editha bayensis*) from the highly protected and well-studied Jasper Ridge site adjacent to Stanford's Center for Conservation Biology was discussed (McGarrahan 1997). Since at least 1934 the 485-ha preserve has been used for biological studies. The preserve is now a remnant—an island in a semisuburban sea and one of only a few places in northern California where one can still find native grasses on stony, inhospitable, serpentine soils. In 1960, the butterflies lived in three distinct populations on Jasper Ridge. By 1987, the butterfly was listed as an endangered species. Fortunately, another larger population was discovered elsewhere, so Stanford biologists took the opportunity to record the local extirpation of the butterfly.

The discontinuances of cattle grazing on Jasper Ridge altered the composition of the plant community the butterfly depended on. Other stochastic events such as inadvertent malathion® treatment in 1981 and local pesticide use added to the death toll. Ultimately, the last remaining butterflies were restricted to a mere 2 ha. Purely stochastic bad weather and lack of topographic diversity sealed their fate. The plants upon which the butterfly larvae depended withered and died early on warmer slopes, and later on cooler slopes. Without adequate habitat variations to choose from, the butterflies died off. Even noted biologist Paul Ehrlich, who had studied this species *in situ* since 1960, suspected there existed several large patches of suitable habitat that could be effectively utilized by the butterflies. He was wrong. Ehrlich concluded that more than total area matters. Topographic diversity of the habitat was an absolutely critical factor, as was the timing of the drying of the plants. The vagaries of local weather patterns and the lack of habitat heterogeneity caused by the cessation of grazing by herbivores led to the extirpation of the Bay checkerspot butterfly on Jasper Ridge.

Jasper Ridge might have been labeled a "hot spot" for local diversity. The site was fully protected and well studied. Indeed, more was known of the butterfly and its habitat than most species and places. The theme that saving so-called hot spots will preserve biodiversity is simply preposterous. Setting aside vignettes of biodiversity such as national parks and preserves will also fail. Inevitably, context matters. Episodic events such as inadvertent insect spraying, catastrophic summer thunderstorms, or local hard freezes happen, not with regular frequency, not predictably, but inevitably. The extent of the unexpected event is proportional to the time span between such events. Small unexpected events happen more frequently than large unexpected events. Invariably, however, the unanticipated occurs. Indeed, the only certainty is that the unexpected event will occur. Without migratory connectivity and landscape heterogeneity long-term protection of species is not possible.

A Top Predator in a Highly Fragmented Human-Dominated Landscape

The work of one of the authors (JGS) on guignas (*Oncifelis guigna*), a small South American felid, showed that guignas were typical wild cats. Males occupied home ranges that overlapped several females, and female ranges were much smaller and entirely contained in male ranges. Males traveled greater distances to defend their territories. In the human-dominated, highly fragmented landscape of Isla Grande de Chiloé, Chile, this need to travel greater distances inevitably brought male guignas into contact with domestic fowl, dogs, and cats. On rare occasions such encounters were fatal to guigna males. Guigna females avoided such encounters entirely. But such effects of fragmentation are just that—landscape effects.

To understand the ecology of this effect on the landscape (e.g., on the space) requires far more time and effort. The question that must be answered is, what effect do guignas have on the ecology of the landscape, and how does fragmentation change the ecology through its effect on guignas? Guignas evolved in temperate rainforests of the southern cone of South America. They are top carnivores in some of the regions they occupy such as on Isla Grande de Chiloé. I suspect their diet of birds, mammals, lizards, and insects has a top-down effect on the ecology of these forests. I have not yet succeeded in elucidating this effect, however. I do know that guignas use corridors, and landscape connectivity is crucial if guignas are to inhabit an area. With the changes that are occurring on the island there might not be sufficient time to answer this fundamental landscape question. Recall Simberloff (1998):

> The recognition that some ecosystems have keystone species whose activities govern the well-being of many other species suggests an approach that may unite the best features of single-species and ecosystem management. If we can identify keystone species and the mechanisms that cause them to have such wide-ranging impacts, we would almost certainly derive information on the functioning of the entire ecosystem that would be useful in its management.

Simberloff's advice offers a valuable approach to the study of the ecology of landscapes when such a study is undertaken as Johnston (1995) and Bowyer et al. (1997) suggest. As might be suspected, insight into how organisms, especially top carnivores, influence landscape ecology can be built upon the thousands of natural history studies of single organisms already completed. We are not suggesting a new approach to the study of ecology, but rather asking researchers to extend those studies by considering the top-down effects the subject organisms have. One way to prevent further degradation of the biological resources of the earth is to make the case that these organisms, especially those threatened or endangered with extinction, have a measurable impact on the ecology of landscapes. Our challenge is to quantify this impact that may, in fact, only be measurable in ecological time.

Considering mobile organisms in a landscape context enables a major step toward the conservation of biota in general. Certainly this research thrust will add impetus and direction to the discipline of landscape ecology and highlight contributions to biological conservation where conservation should take place — across the landscape.

EXERCISES

5.1 Summarize the current understanding of the disappearance of human societies inhabiting Mesopotamia, the so-called Cradle of Civilization, the Mayan Empire, and early Easter Island. See Ponting (1992).

5.2 Compare the landscape effects of subsidized cattle grazing in the American West with cattle grazing on private lands in the east.

5.3 What is GAP analysis? Discuss its pros and cons.

6

Landscape and Edge Effects on Population Dynamics: Approaches and Examples

Lennart Hansson

CONTENTS

Introduction

Until the early 1980s, population dynamics were almost always modeled, conceptually or mathematically, for a homogeneous area without any edge. However, most habitat patches are small, particularly with regard to wide-

1-56670-368-9/00/$0.00+$.50
© 2000 by CRC Press LLC

ranging vertebrates, and edge effects are indeed common. It is surprising that field ecologists did not object to this unrealistic representation of nature until 10 to 15 years ago. There was a gradual change in the late 1980s and a more pronounced one in the 1990s, particularly with the re-apprehension of the metapopulation approach. And during the 1990s the theoreticians have again found a new playground, now starting to model effects of environmental heterogeneity. However, the early observations about "suprahabitat" or landscape effects (Levins 1969; Anderson 1970; Hansson 1977; Wegner and Merriam 1979) were usually not sympathetic to traditional theory. In this chapter I will look at dynamics in heterogeneous areas (usually the landscape scale, as discussed later, with two or several ecosystems) from a field biologist's rather than a theorist's point of view.

Landscape composition has various influences on populations of mobile organisms. So far, mainly animals have been considered, but effects on plant propagules ought also to be examined. Furthermore, nonrandomly distributed mobile herbivores may have pronounced local effects on sedentary plant populations. The use of landscapes by fungi or microorganisms has hardly been considered at all. Thus, this discussion will mainly be limited to animals.

Environmental heterogeneity can have a multitude of effects on populations. A list of sensitive population attributes, not always mutually independent, may include:

a. Distributions. Population may shift from uniform to clumped distributions depending on habitat patch size or juxtaposition.

b. Persistence. The longevity of local populations is dependent on the size of the inhabited area, dispersal routes and distances to other populations.

c. Type of dynamics. Important predators or pathogens, or the physical environment, may affect dynamics differently in various spatial contexts.

d. Regulation or density level. Populations may lose individuals at edges, affecting mean density level or setting an equilibrium.

e. Habitat overflow. Certain habitats allow higher recruitment than others, causing population expansion or source–sink dynamics in a landscape.

f. Habitat interdependence. Individuals from one habitat may glean necessary or additional resources in a neighboring habitat whereby population growth rates are impacted. Habitat juxtaposition may be necessary or beneficial.

g. Connectivity between habitat patches will affect population sizes and existence.

These landscape effects are interwoven for most populations. Here, I will present types of populations that are affected by environmental heterogeneity, and provide some examples, mainly from research at Uppsala, Sweden, or in Scandinavia generally. The ecological setting is boreal to temperate environments. I will finally try to make generalizations regarding common effects of environmental heterogeneity, evolution of landscape use, and the future development of landscape ecology.

Metapopulations

The idea of random extinction and colonization of subpopulations in isolated habitat patches in the theory of metapopulation dynamics by Levins (1969, 1970) was a pronounced break with contemporary theory, but there was a lapse of some 20 years before it was fairly generally accepted. It was, thus, the first approach to population dynamics in a heterogeneous environment (Figure 6.1). Now it is at the center of much conservation theory (Hanski 1994; Gilpin and Hanski 1995). However, in order to be general it contains clearly unrealistic components as similar-sized patches at similar distances. Its predicted equilibrium populations have actually been observed in fairly few instances (Harrison 1991). Indeed, many cases of metapopulations, the concept taken very widely, have turned out to be satellite or sink populations to larger source areas (Pulliam 1988) or declining populations, particularly of endangered species (Hanski 1996).

I will demonstrate some problems of applying the metapopulation theory to any subdivided population, using as an example the pool frog studied by Sjögren Gulve (1991, 1994) and Sjögren Gulve and Ray (1996) close to Uppsala in south-central Sweden. At its northern distribution border, it occurs in many separate subpopulations, breeding in pools isolated from the Baltic by land uplift. The total number of permanently or temporarily occupied pool populations was defined as a metapopulation by Sjögren Gulve. However, the pools were not identical, as in Levin´s model, and differed in several recognizable features: occupied pools had higher water temperature in spring and were close to other occupied pools. There were both deterministic and more "stochastic" extinctions, the former being due to pool succession or drainage. However, also the latter type of extinctions were predictable in the sense that they occurred close to the primary deterministic extinctions. Thus, dispersal was a critical element in the metapopulation system, declining with distance from source populations. Dispersing frogs were moving through forests between the pools and also overwintered in moist forest sites. This dispersal and wintering was evidently negatively affected by large-scale drainage and clear-cutting, creating an environment too dry for successful dispersal (Sjögren Gulve and Ray 1996). Thus, although this particular metapopulation appeared to be at equilibrium (Sjögren 1991), the colonizations and

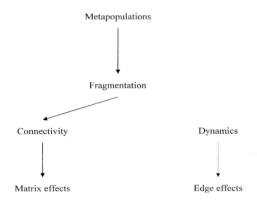

FIGURE 6.1
Particularly emphasized subjects in the short history of a landscape ecology centered on populations. There is a conceptual line running from metapopulations to matrix effects over fragmentation and connectivity while dynamics and edge effects have been made more separate subjects. It should be possible to come to a joint understanding of matrix and edge effects.

extinctions were not random processes, but were strongly affected by the matrix between habitat patches.

I suspect that similar conditions occur within most "metapopulations" (see, e.g., Kindvall 1996 below) and that effects from the matrix have to be considered for most subdivided populations. The metapopulation concept is strongly connected to purely stochastic processes and if such circumstances do not apply, the term "subdivided populations" and identifications of matrix effects might be preferable.

Types of Dynamics

Variations in dynamics include equilibrium populations, outbreaks, cyclic, and chaotic populations (e.g., Hassel and May 1990). Populations may also move towards extinction. Populations in different parts of a geographical range can show different types of dynamics and local populations can change from one type of dynamics to another. Most examples come from folivorous insects (Berryman 1981), but similar variation has been observed among folivorous mammals (Pimm and Redfern 1988). Both regional and temporal variation may be due to landscape composition.

I will first discuss landscape effects on the regionally different fluctuation patterns in Scandinavian small rodents. A geographical gradient in fluctuation patterns in Scandinavia is well established (Hansson 1971; Hansson and Henttonen 1985; Hanski et al. 1993; Turchin 1993; Björnstad et al. 1995), with more heavily fluctuating populations in the north. At population peaks in northern Scandinavia a wider spectrum of habitats is also utilized than in south Scandinavia (and in central Europe). The level of fluctuation is positively related to the amount of snow cover (Hansson and Henttonen 1985). The explanation suggested is that the snow prevents predation by large generalistic predators by isolating the small rodents below the snow and by diminishing numbers of alternative prey. In areas poor in snow, generalists may stabilize rodent numbers by switching between prey species. In snow-rich areas, predators living under the snow get an advantage, but they consist only of specialized mustelids that will overexploit the rodent populations and cause the pronounced population cycles in voles that are so typical of northern snowy regions (Korpimäki et al. 1991). The snow is thus the factor that is causing homogenization of various landscapes for different lengths of time. These observations and explanations have changed the way people have been looking at rodent dynamics (Stenseth and Ims 1993) and, although not globally accepted, this hypothesis is supported by various simulation models (Hanski et al. 1991; Björnstad et al. 1995; Hanski and Korpimäki 1995; Hanski and Turchin, unpublished).

Temporal variations in population fluctuations have also occurred in Scandinavia (Hansson 1992, 1994; Hörnfeldt 1994; Lindström and Hörnfeldt 1994; Hanski and Henttonen 1996). Small rodents were strongly cyclic in central Scandinavia in the 1960s and 1970s, but gradually lost their high degree of cyclicity in the 1980s, and these latter populations appeared almost stable with only seasonal density variations in the early 1990s. Conversely to the geographical gradient, there is not any generally accepted explanation of the temporal change, but the following one is in agreement with recognizable changes in the landscape.

Extensive clear-cutting started in the 1950s in central Scandinavian forests, perhaps particularly in central-northern Sweden. The most productive forests were cut first, and the clear-cuts were then really huge. At the same time, a lot of agricultural land was abandoned in the same region. Early successional, grassy patches on forest ground are prime habitats for *Microtus* voles, as are fairly early phases of overgrowing farmland (Stenseth and Lidicker 1992). From the 1970s onwards smaller clear-cuts were taken up on less productive land, old clear-cuts were reforested and no longer of any use to *Microtus* (Hansson 1994), while abandoned fields went into stages dominated by coarse grasses that are less profitable to *Microtus*. Thus, there were large and close patches of sheltering and nutrient-rich vegetation in the 1960s and dispersal between patches was simple. High reproduction and population growth rates then prolonged the time before weasels could overtake the *Microtus* populations, but when they finally did, they depressed the voles to very low-density levels. In the 1990s, productive patches were small and distant, and small or transient, slow-growing *Microtus* populations were less able to colonize new patches. Weasel populations did not reach as high densities as previously (Hansson, unpublished) and generalistic predators may have had great impact during the summer dispersal. Many vole populations now also declined during summer. *Clethrionomys* voles in adjoining forests, which have lower population growth rates, may have fluctuated due to switching between prey species by weasels and other rodent predators (Heikkilä et al. 1994). However, *Clethrionomys* voles also prefer the most luxuriant forests that were cut already early on.

Similar observations and suggestions were made for disappearance of vole cycles at the well-studied Wytham Wood near Oxford by Richards (1985). For an alternative, or rather complementary, explanation involving chaotic dynamics, but also interactions between clear-cut–living *Microtus* and forest-living *Clethrionomys* via predators, see Hanski and Henttonen (1996).

Similar large geographical and temporal variation has been observed in the population dynamics of insects, e.g., the winter moth *Epirrita autumnata*, showing pronounced population outbreaks in certain areas, but not in others, in northern Fennoscandia and never in the European Alps (Tenow and Nilssen 1990; Berryman 1996). Evidently, accumulation of cold air in valley bottoms kills the insects' eggs while the insects may thrive on valley slopes and there exhibit pronounced population cycles.

Fragmentation of, e.g., forests may or may not cause declining populations and extinctions, depending on movement characteristics of focal organisms. There may be only a statistical sampling effect or true isolation effects (Andrén 1994, 1996). Dispersal ability is thus very important for the final outcome. The European red squirrel displays isolation effects for forest patches surrounded by agricultural fields, but not when surrounded by clear-cuts (Andrén and Delin 1994, Delin and Andrén 1997); varying behavioral reactions to the matrix type are evidently crucial.

Observations of population extinctions in heavily fragmented habitats are commonplace. In Scandinavia, the boreal forest was fragmented by forest fires in the pristine state, but moist or wet forests did not burn or burned very seldom (Hansson 1997). Many species inhabiting such moist forests show little dispersal. Species adapted to these fire refugia are therefore very sensitive to modern forest management with extensive clear-cutting in smaller or larger blocks. Examples of such organisms with populations at present moving towards extinction are several lichen species depending on a humid environment, e.g., the large lichen *Usnea longissima* (Esseen et al. 1997).

Spatial Distribution

The distribution of individuals in a population is commonly expressed as even, random, or clumped, usually after statistical tests in relation to a Poisson distribution. These distributions may be affected by landscape composition, at least when populations fluctuate strongly.

Small rodents on clear-cuts in fairly stable south Scandinavian populations show a pronounced clumpiness while cyclic small rodent populations in north Scandinavia demonstrate generally even or random distributions (Hansson 1990). However, in the southern region there were areas with deviating distributions. Snow depth explained a large proportion of the variation between the southern and northern populations (Hansson 1989). The low level of clumping in the north is supposed to be related to an easy dispersal under a landscape-wide snow cover (Hansson 1992). Even in noncyclic populations exposed to heavy predation, dispersal may be more or less easy, depending on the density or structure of the vegetation.

These animals still change habitats or microhabitats during a population cycle. On central Swedish clear-cuts, the vole *M. agrestis* moved from wet parts of clearcuts, with dense grass cover at the peak and early decline to boulder fields with less food at the same time as weasels (*Mustela nivalis*) invaded the grassy parts (Hansson, unpublished). Similar but less pronounced changes were observed in the more mobile vole *Clethrionomys glareolus*. The occurrence of alternative microhabitats evidently prolongs the decline phase.

These observations imply that folivorous species in homogeneous habitats or landscapes may show even distributions at large density variations and that the dynamics may be still more violent the more homogeneous an area or habitat. Heterogeneity and clumped distributions may be general indicators of population stability, however a prudent proposition to be examined in other systems.

Effects of Habitat Juxtaposition

Population Effects

Many species are either dependent on two or more habitats (landscape complementation, Dunning et al. 1992) or favored by more than one habitat (landscape supplementation, Dunning et al. 1992), both conditions being affected by the level of connectivity through the intervening matrix (Taylor et al. 1993). These dependencies are usually supposed to work close in time, daily, seasonally, or at least annually. Spatial scales may vary and examples will be provided later. However, long-term effects would also be possible, with far-reaching implications for conservation. The proximity of such habitats might affect the type of dynamics within one or several habitat types. Interspecific interaction may also influence dynamics within adjacent habitats. I will here exemplify these additional effects.

A bush cricket, *Metrioptera bicolor*, studied by Kindvall (1995, 1996) in southern Sweden occurs in subdivided populations in grasslands, surrounded by dry pine forest. During most years this species performs equally well on various types of grasslands, but during one particular year (1992), with a long-lasting drought, the crickets survived less well on short-grass grassland and even moved into the forest during the main drought and survived in the shade there. Local extinctions would have occurred if the main habitat had not been surrounded by drought-tolerant vegetation. Similar observations of spatially varying weather effects have been made on caterpillars of an American butterfly *Euphydryas editha* associated with serpentine grasslands (Dobkin et al. 1987).

For the cricket *M. bicolor*, habitat heterogeneity, measured as size variability of various microhabitats in a sampling plot, was negatively correlated with population variability (Kindvall 1996). Populations in some of the most homogeneous plots went extinct during a study period. This is one further example that populations in large and fairly homogeneous habitats often are unstable, either with regard to fluctuation patterns or to local persistence.

Adjacent habitats may show multiannual variations in quality for a certain species. A fairly extreme case is the mast seeding of certain deciduous forest groves that are still retained from earlier extensive hardwood forests within the present conifer forest landscape in south central Sweden. The extensive conifer forests support only very sparse populations of wood mice (*Apodemus flavicollis* and *A. sylvaticus* (Hansson 1997). At mast seeding in the deciduous groves there are population outbreaks of granivorous rodents like these *Apodemus* species, but the magnitude of the increase depends on the densities in the surrounding conifer forests (Hansson 1997), outbreaks in extensive deciduous forests being much more rapid and reaching higher densities (Pucek et al. 1993). Similar conditions have been observed at artificial feeding with seeds in small patches (e.g., Hansson 1971). The population outbreak in the groves by various small mammal species caused at least one rodent species, *C. glareolus*, to affect surrounding conifer forest densities by dispersal after peak food availability.

Community Interactions

Under the mast-related population outbreaks, pairs of congeneric species segregate in space, *A. flavicollis* dominating in the deciduous groves and *A. sylvaticus* increasing for a shorter period in adjoining conifer forests or other adjacent environments (Hoffmeyer and Hansson 1974; Hansson 1997). Similar observations were made for the insectivorous shrews *Sorex areaneus* (deciduous groves) and *S. minutus* (conifer forest), probably relying on seed insects (Hansson 1997). In periods without mast production, little evidence of competition is detected (e.g., Hansson 1978).

If a prey or host species to a predator or pathogen, respectively, is affected by landscape composition then the predator or disease will also depend on this landscape. If a predator, pollinator, etc. depends on two or more species of prey, plants, etc. in different habitats then landscape composition will again influence predator, etc. distributions and probably also numbers. Finally, if effects of predation are strong, then alternative prey in another habitat may be affected at bottlenecks in the staple food. The later conditions are often discussed under the heading of "apparent competition" (Holt 1984).

A case of possible apparent competition (i.e., changing prey) in various habitats is modeled by Hanski and Henttonen (1996) and may explain certain features of Scandinavian vole cycles. For instance, this model may be related to the synchronous dynamics of one rodent species (*Microtus oeconomus*) living on small grassland patches in northern Finland and heavily preyed upon by stoats and another rodent species (*C. rutilus*) living in adjoining extensive forests with initially little predation (Heikkilä et al. 1994). The grey squirrel (*Sciurus carolinensis*) has invaded Great Britain and "driven away" the red squirrel (*S. vulgaris*) in spite of different habitats for these two species. However, a common disease, the grey squirrel being resistant, may possibly have been the agent (Reynolds 1985). Similar disease conditions appear to mediate "competition" effects of the introduced American crayfish *Pacifastacus leniusculus* on the indigeneous species *Astacus fluviatilis* in Scandinavia (Gydemo 1996).

Behavior at Edges

Most habitat patches are generally small, particularly with regard to wide-ranging vertebrates, and edge effects are indeed common. Different species respond in particular ways to a certain edge and there has been a separation between soft and hard edges (Stamps et al. 1987). Many new processes occur at edges and others are intensified there (Hansson 1994), and populations of various species may respond differently to these changes. There has been an emphasis on negative effects (climate, predators, etc.) on populations of forest interior species (e.g., Temple and Cary 1988), while the use of edges and surrounding habitats by ecotonal species has been considered less.

Different movements and dispersal rates of various population categories cause different population compositions at edges and in interior habitat (Gliwicz 1989; Hansson 1997). At a decline in food resources, many granivorous rodents remained at the edges, exploiting food both in the deciduous grove and surrounding conifer forest. In this case, the edges provided the best out of two worlds (or habitats) for generalized species. However, most specialized species disappear from edges, but remain in larger tracts of interior forests (Hansson 1983, 1994 for various bird and mammal species). Similar observations have been made for North American tropical migrant birds (Whitcomb et al. 1981). Edges appear generally to be terminators for specialist species, but refugia for generalist species.

Animal reactions to edges are, however, not constant. In the study on the cricket *Metrioptera bicolor*, Kindvall (1995) found the imagoes to cross grassland–forest borders extensively in a prolonged drought, a behavior not seen under normal weather conditions. These movements were evidently due to physiological or psychological changes. Similarly, wood mice invaded mast-seeding groves due to the rich food supply.

Effects at Various Spatial Scales

Landscape Complementation or Supplementation at Short Distance

A species may need different resources during daily life or within short time periods. These requirements can be fulfilled within a habitat or by movements between adjoining or nearby habitats. It may sometimes be difficult to distinguish between habitat and landscape effects, as the following examples demonstrate.

Passerine birds in the hemiboreal zone show higher mean population numbers in mixed conifer–deciduous forests than in pure deciduous or conifer forests (Nilsson 1979). Evidently they require food resources, particularly insects and seeds, from the deciduous trees and shelter against predation in the dense but dark and insect-poor conifers. Conifers and deciduous trees can grow intermingled (one habitat) or close by (a landscape?).

The Siberian tit (*Parus cinctus*) of northern Finland prefers habitats with dead hollow trees, large coniferous trees and insect-rich birches. Nesting success was considerably higher when there was access to these features, even within moderately managed forests (Virkkala 1990). During the fledgling period the tits move to habitats with more birches, and insects, than in the nesting period (Virkkala 1991). As a result of removal of old birches in forestry, the Siberian tit numbers are now declining regionally.

Scandinavian old-growth forests provide habitats for a high biomass of foliose and pendulous lichens. These lichens are used as shelter by various insects that are preyed upon by spiders, which in turn are preyed upon by birds (Pettersson et al. 1995; Gunnarsson 1996). With each higher trophic level, a wider spatial scale is affected by old-growth remnants.

Effects at the True Landscape Scale

Complementation and supplementation are common within the 5- × 5-km scale that is proposed as the genuine landscape scale by Forman and Godron (1986). Most examples come from larger vertebrates, but few other organisms have been examined very carefully in this respect.

Scandinavian roe deer forage on fields or clear-cuts, but seek shelter in forests, particularly at edges (Hansson 1994). Conditions appear very similar for North American white-tailed deer (Alverson et al. 1988). Corvids such as European crows, rooks, and jackdaws feed on fields in winter, but roost in forest groves or city parks (Jonsson 1992).

Such dependencies may strongly affect population numbers; jackdaws can occur in tens of thousands at night in a small town surrounded by open fields (e.g., Uppsala, Hansson, personal observation). Similarly, deer such as moose and roe deer can increase to such numbers by affluent food within daily ranges at nonintensive farming or extensive clear-cutting that they cause damage in forests at crucial seasons (Cederlund and Bergström 1996). Inefficient use by humans of natural resources coupled to sheltering environments appears more generally to provoke intense population growth of generalist species, such as crows and rats (Golley et al. 1975).

Effects at Larger Spatial Scales

For species with great mobility, effects may occur on still larger scales. It is probably possible to separate effects on a regional scale (cf. Forman 1995) and a more or less global scale. The regional scale may be set to around 500×500 km while "global" effects would be over still larger distances.

Regional effect may occur if, e.g., one habitat is changed so that it produces a large surplus of individuals (a source area) that swamp many other habitats. Such conditions will mainly occur for generalist species. Examples are generalistic passerine birds that have been favored by forestry in south Finland and now occur in high densities also in less modified forests and also in nature reserves all over the country (Haila et al. 1979; Helle 1986, Väisänen et al. 1986).

Global dependencies are most obvious for migrating birds, fishes, and butterflies, to mention the most well-known groups. Examples from the largest scales are also nomadic birds that are specialized with regard to food, but not to habitat. Such species include nutcrackers looking for mast, crossbills eating conifer seed, waxbills depending on berries, and raptors and owls depending on small rodent prey. An extreme example is the snow owl that appears to have a circumpolar range in its search for lemming peaks. However, movements of the slenderbilled nutcracker and nuthatches from Siberia to Scandinavia are almost as impressive. Many such irruptive individuals survive on food in distant places and then move back; others start breeding in the new environment (Korpimäki et al. 1987; Alerstam 1988).

A more recent type of global effect is the invasion of whole continents by preadapted species, a phenomenon that has occurred both in Europe and in America. Examples are the muskrat, released as a fur animal in Russia and now invading Scandinavia (Danell 1977) and the starling, brought to New England by sentimental immigrants and now occurring over most of North America (Long 1981).

Some Limited Generalizations

From the previous survey it might be possible to make a few preliminary generalizations. Heterogeneity evidently improves persistence. This appears particularly important for fugitive or competitively inferior species (Hanski 1995). Heterogeneity provides for more stable dynamics. This is most clear for species that are sensitive to disturbance or inclement weather. It will also apply to predation-sensitive species, providing enemy-free space (Jeffries and Lawton 1984). Heterogeneity leads to large numbers in generalist species that are able to temporarily exploit resources in new surroundings. Habitat edges are important for persistence and recovery for strongly fluctuating populations, particularly when involved in strong trophic interactions, and for most populations in situations of habitat-dependent fluctuations in resources.

It might also be possible to talk about an impact of heterogeneity on "population regulation" in a more general sense, with the implication that populations are kept on a fairly even density level. A large number of suitable habitat patches will permit escape in space and stability in fugitive species. Extensive landscape complementation will permit stable levels of species requiring two or more habitats in their life cycles. Removal of individuals of specialized species at habitat edges will work in a regulatory manner in the more strict meaning of "regulation."

Evolutionary and Historical Background

Why have dependencies on heterogeneous landscapes developed for individuals or populations? I see several reasons and only some of them are based on evolutionary adaptations.

The most basic reason may be competition among plants, perhaps even before animals had evolved. Such competition caused adaptations to various topographic and soil sites, producing a clumping of vegetation, both regarding chemical composition (e.g., nutritional value) and physiognomy. This ecological clumping was enhanced by the development of specialized herbivores or parasites and forced organisms that developed or adapted later to accommodate to this heterogeneity. Nonspecialized herbivores had to look for food in certain patches and shelter in others. This pattern might have been reinforced by efficient generalistic predators. These lines of evolution may have occurred during overlapping time scales.

In seasonal environments there is great temporal variation in resources, particularly food supply for animals. Mobile organisms have always tried to exploit extra resources to increase individual fitness—the reason to be mobile! Habitats vary in productivity and generalistic species will track temporal and spatial patterns in availability of resources. This exploitation can appear spread in time (migrant birds) or space (deer coming out of forests to feed on crop fields), but usually both dimensions are involved. Furthermore, during more recent time the exploitation of natural resources by man has resulted in much spill; generalist animals are now partly utilizing habitats such as cereal stubbles, abandoned meadows, and forest clear-cuts that did not exist previously.

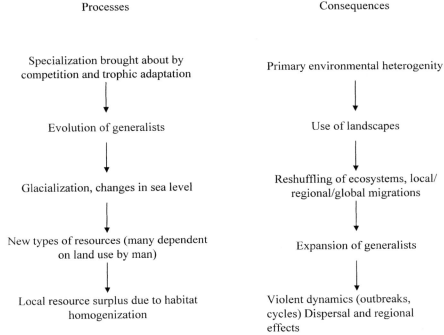

Processes   Consequences

Specialization brought about by   Primary environmental heterogenity
competition and trophic adaptation

Evolution of generalists   Use of landscapes

Glacialization, changes in sea level   Reshuffling of ecosystems, local/
                                             regional/global migrations

New types of resources (many dependent   Expansion of generalists
on land use by man)

Local resource surplus due to habitat   Violent dynamics (outbreaks,
homogenization   cycles) Dispersal and regional
                                             effects

FIGURE 6.2
Evoluntionary and historical development of landscape utilization: a conceptual model.

Adaptations provide for yesterday; the environment of today is changing more or less rapidly. In northern areas there has been a gradual change since the last glaciation and species immigrating there with shorter or longer delay are taking over habitats and landscapes that often did not exist in their evolutionary history. During recent centuries man has changed this environment still more and many species now either go extinct or increase in numbers. Man is generally homogenizing the rural environments (creating large blocks of cereals, even-aged forests, etc.) and animals specialized on rare habitat features or specialized for utilizing several nearby habitats in their life cycle are now encountering problems. On the other hand, a few preadapted specialists and many generalized species prosper, sometimes to an extent that leads to population outbreaks or density cycles (Figure 6.2).

Heterogeneity or Landscape Ecology?

There is a continuing discussion about the scientific content of landscape ecology; some authors see it as a new discipline covering all aspects of environmental heterogeneity (e.g., Wiens et al. 1993). However, nature is heterogeneous at all spatial scales, even within patches that we recognize as a "homogeneous" habitat for a certain species. I would instead define landscape ecology as a consideration of structures and processes at one particular spatial scale. Thus, heterogeneity can be examined within habitats or ecosystems, within landscapes (in the order of 5- × 5-km areas), regionally, and globally (Figure 6.3). However, I believe that the landscape scale is particularly important in man-affected environments as it strongly associates to the scale that is employed in human planning and management.

Thus, I do not regard landscape ecology as a new science. Instead, it relates to a certain scale and the interactions that occur at that scale. Population dynamics at the landscape scale treat the processes previously outlined for one or a few populations at varying temporal scales, however, usually only for one or a few years. The landscape scale can also be related to shorter or longer time periods and larger organismic scales as whole communities or ecosystems.

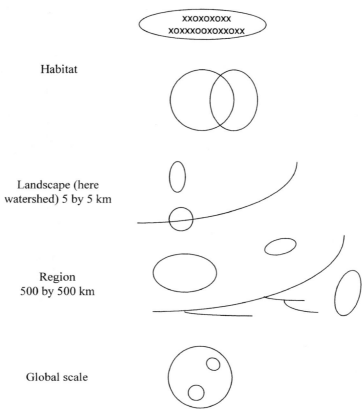

Habitat

Landscape (here
watershed) 5 by 5 km

Region
500 by 500 km

Global scale

FIGURE 6.3
Spatial scales for population processes. See text for detailed explanation.

How to Test Landscape Effects

Problems in landscape ecology are often of local concern and usually they cannot be backed up by any general theory, neither do they provide any results to be easily generalized. Furthermore, the orthodox way to test hypotheses, by manipulative experiments, is usually closed: the scale is too large and too many species or habitats are involved. As an alternative, ecological model systems (EMS) have been proposed for testing more general heterogeneity effects, particularly on populations, and provide empirical generalizations (Ims and Stenseth 1989; Ims et al. 1993; Wiens et al. 1993); small-scale studies are extrapolated to solve large-scale problems, even including an upscaling over several orders of magnitude. The applicability of this approach has to be proven; however, it should be realized that the complexity of ecological systems is commonly scale dependent: laboratory experiments exclude weather effects, field enclosures exclude predation, and time limitation excludes "catastrophic" events.

Another way of "testing" proposals in landscape ecology is by mathematical modeling and simulation (e.g., Hanski et al. 1991; Hanski and Henttonen 1996). Such models and simulation may examine the internal consistency of a hypothesis and provide correlations between model output and measured field variables. However, simulation models are often too complex to permit decisive evaluation of the effects of single variables.

What remains are descriptive studies on a multitude of species and environments. In that way inductive generalizations may be possible. But studies on the landscape level are problematic and expensive: we probably will have very few basic and prudent studies in the required spatial and temporal scale. However, there is a possibility in this direction that has not been exploited, i.e., the common studies on endangered species, pest, and game species. This applied work could habitually be directed towards the landscape approaches in order to collect as much useful information as possible. Furthermore, this also offers the possibility to perform management experiments (adaptive management, Walters and Holling 1990), that may not fulfill all requirements of classic experimental tests, but still partly examine the relevance of general or context-specific hypotheses.

Landscape Ecology and Conservation

Biodiversity and conservation are now probably the most common objects for ecological research, and much landscape ecology has been connected to such applied work. Outcome of such studies may thus refine principle and generalizations within the landscape scale of ecology. I will therefore end this treatise on populations in landscape contexts with a few reflections on the relevance of this approach to conservation.

Conservation now displays species and ecosystem contrasts. For pristine nature there might only be one common solution: conserve original functional landscapes. However, we have to delimit such functional landscapes both for mobile species and for ecosystem processes. In a rapidly changing world, the recipe for feasible last-moment actions will probably often be: save as much as possible, but always provide buffers against man-modified environments—an important landscape consideration!

For managed areas, the scientific problems are actually greater and we really first have to decide what to preserve: I believe that we must set a time frame for the state of the environment to be preserved and clarify all links to other (often prior) human environments for the focal ecosystems or species. We need to reconstruct and maintain entire original landscapes! The emphasis should be on crucial ecosystems and specialized species in workable landscape contexts.

In both cases we have to change the present preoccupation with preservation of isolated populations or ecosystems to a conservation of functional landscapes as the main spatial unit for applied ecology. However, nature works on even larger scales and saved or restored landscapes will not retain or protect everything!

Acknowledgments

Comments from Larry D. Harris and Jim Sanderson were highly appreciated. My research as reported here has been supported by the Swedish Council for Natural Sciences and by the Swedish Agricultural and Forestry Research Council.

Part III

Landscape Theory and Practice

7

The Re-Membered Landscape

Larry D. Harris and James Sanderson

CONTENTS

How did the composite set of ecological processes get so out of balance so as to produce the deranged, dysfunctional, dismembered landscapes we have today? We cannot summarize the course of human history here. Humans have proven to be nearly infinitely adaptable and to accept the present as the way things have always been. That is, changes made by humans occur so rapidly that they become the *status quo* in short order. For instance, how long has the Glen Canyon Dam blocked the waters flowing into the Grand Canyon? For much of the U.S. population the answer is "the dam has always been there."

A complacent acceptance has quieted what should be outrage. We are disappointed not to be able to see walruses along the Northwest Atlantic shores from Cape Cod to Greenland. We desire their return. The problem is not that walrus populations cannot be restored; the problem is that widespread ignorance of

1-56670-368-9/00/$0.00+$.50
© 2000 by CRC Press LLC

the previous existence of walruses prevails. History, and not just ancient history, is being forgotten.

Humans have now developed technology sufficient to alter nearly all processes affecting landscapes—deliberately and otherwise. There are proposals to use nuclear weapons to deflect potential earth impactors long before they are even close to the planet. Humans have altered the chemical composition of the atmosphere, warming the global climate. Can we now alter the ocean currents with such changes? Humans realized the value of movement corridors as the need for communication and trade increased. The U.S. Highway Act created the most extensive and complex movement corridors on earth for the benefit of humans. With few exceptions this has proved disastrous for ecological processes, especially wildlife movement. River courses have been altered and repeated attempts have been made to control the flow of the great rivers of the world including the Yangtze River in China. The disastrous negative effects on huge delta ecosystems such as the Mississippi delta and the Nile delta have become appreciated and apparently been relegated to the dustbin of history. The complete alteration of the regional climate surrounding the Aral Sea in Asia and the nearly total degradation of land- and seascape processes was achieved by humans in less than 30 years. Cattle have overgrazed the western U.S. from Mexico to Canada, at a cost that can scarcely be calculated and shouldered by the American public for the benefit of a comparatively few citizens.

These changes in and of themselves are remarkable achievements. Moreover, they are cumulative, rarely canceling previous alterations. Recognition of the negative effects of these changes and numerous others wrought by humans has lead to a few restoration efforts. However, restoring nature has proved elusive.

The U.S. Department of Agriculture estimates that restoration and creation projects have added more than 400,000 ha of fresh- and saltwater wetlands in the U.S. since 1982. In 1998, the Clinton Administration called for the creation of another 80,000 ha of new wetlands each year for the next decade. The goals of the project were that the wetlands should be "functionally equivalent" to undisturbed, natural wetlands. But can nature be recreated? Experiments are underway now that compare recreated wetlands and their nearby natural systems (Malakoff 1998).

Landscape fragmentation continues around the world, creating yet more dismembered landscapes. Once landscapes are dismembered other ecological processes such as invasions of "weedy species" occur, further changing natural ecological processes. Recently, Wahlberg et al. (1996) created a model to predict the occurrence of endangered species in fragmented landscapes. The modeling approach was presumed to be a practical tool for the study and conservation of species in highly fragmented landscapes. In the model, the probability of local extinction was determined by the size of the habitat patch. Isolation from occupied patches and the size of the patches determined the probability of colonization of an empty patch. Empirical data to support the model came from studies of the Glanville fritillary butterfly (*Melitaea cinxia*).

The model was then used to predict the patch occupancy of the false heath fritillary butterfly (*M. diamina*). The benefits of such a model are numerous. Can such a model be useful for all species?

The size and isolation of the patch were used to determine the presence of butterflies in patches. From our landscape perspective the analysis on the contextual setting of each habitat patch is equally as important as the patch itself. That is, if an isolated habitat patch was considered close to occupied patches by some distance metric, then the isolated path would, with high probability, be occupied. Linear distance, however, is a poor metric to measure isolation. To appreciate this, suppose, for instance, that all isolated patches within 100 m of each other were occupied and that one isolated fragment was separated by a mere 50 m and a six-lane superhighway from these occupied patches. The model would predict that the isolated patch would be occupied without regard to the *physical barrier* created by the highway. Moreover, between-patch physical distance was assumed to be *invariant* for all species. This suggests that a bald eagle would have just as much difficulty as a mouse in attempting to occupy the habitat patch, presuming both occupied nearby favorable patches.

Isolation of favorable patches can be enhanced by the content of the unfavorable patches (Merriam 1991). An otherwise favorable fragment might lay surrounded by a city as in the case of Central Park in New York. Nearby noise or light pollution might adversely affect birds more than rodents, enabling the latter to colonize patches that no bird would enter, however close a favorable patch might be. Therefore, we must conclude that linear distance is an inadequate currency to measure the colonization ability of a species because different physical barriers to colonization are species dependent. The distance measurement must, at a minimum, be modified to be a "degree of difficulty" measurement that varies between species. A contextual analysis is critical to understanding species distributions. Deciding when a patch is small enough or isolated enough, or determining how wide a corridor must be to enable species movement is not the answer to re-membering fragmented landscapes.

The General Theory of Insular Biogeography

With the previous examples in mind and other well-understood situations we now state four fundamental theories of landscape ecology: Edge Theory, Juxtaposition Theory, External Impact theory, and Corridor Theory. Using these theories we can create a General Theory of Insular Biogeography. Note that these theories do not depend on the size of the fragment, reserve, protected area, or hot spot.

Edge Theory

Generalist species are more likely to be found along edges or ecotones that are avoided by specialist species.

Edges are also where humans are often found. Woodroffe and Ginsberg (1998) recently reported that wide-ranging carnivores were more likely to disappear from protected areas regardless of their population size because such species came into contact with people along reserve edges more frequently. Data on ten carnivores were used to support this conclusion. My own data on *Oncifelis guigna*, a small forest cat, supported and extended this conclusion. My data suggested that male carnivores were more likely to suffer human-caused conflicts than females. This was because males had home ranges that overlapped several female home ranges. Male ranges most often included human homes, and males traveled between females and therefore invariably came into contact with humans, their pets, and domestic fowl. Males more frequently crossed roads, thus risking exposure to domestic dogs. Inevitably, males were more tempted to take domestic fowl, especially free-ranging fowl, than females.

Edges and patches also affect the quality of movement corridors. We know that edges invite invasive species and that nearby unfavorable habitat negatively influences corridors. Would a panther use a linear forest path bisecting a university campus, for instance? The theories we have presented can be applied to the analysis of landscape connectivity and patch influence.

Juxtaposition Theory

Processes within landscape fragments are affected by processes acting in proximate fragments. The impact of the effect extends beyond the boundary of the fragment and depends upon the strength of the process.

Juxtaposition Theory says that processes such as human activities affect other processes acting within fragments. For instance, night light pollution negatively impacts birds in otherwise suitable habitat. Nearby noise or light pollution is a proximate process.

Corridor Theory

Corridors increase population persistence in fragmented landscapes.

Fahrig and Merriam (1985) and Merriam (1991) discussed the role corridors in patchy habitats played in the demographics of small rodents. There were three demographic effects of interpatch dispersal. First, interpatch movement enhanced metapopulation survival. Second, interpatch dispersal supplemented population growth in certain instances. Third, patches where extinction occurred were recolonized. The greater the connectivity between patches, the more likely the metapopulation was likely to persist. Merriam (1991) concluded that connectivity was critical to species long-term survival. But what constitutes connectivity?

Species-specific behavior determines whether or not suitable corridors and landscape connectivity exist. Merriam (1991) noted that the assessment of connectivity must therefore come from species-specific empirical studies. That is, looking at a highly detailed vegetation cover map and quantifying habitat is simply not good enough to determine if landscape connectivity exists for the mobile species considered. Movement behavior must be known.

External Impact Theory

Processes within landscape fragments are affected by external processes whose origin, time of arrival, and strength of impact cannot be known in advance. Nevertheless, with certainty an external process will severely negatively impact natural functioning processes within the landscape fragment.

A hurricane is a natural process that acts episodically. During hurricane season, the probability of an isolated fragment of beach being hit by a hurricane is near zero. However, we can say with total certainty that eventually the isolated beach will be hit. The probability of complete destruction is probably again small; however, given enough time, disaster will occur. Hurricanes, acid rains, or meteorite impacts are examples of processes acting on fragments that are not of proximate origin. That is, these processes originate elsewhere and then travel stochastically, impacting fragments in their path.

These four theories are supported by many examples and have been fundamental research programs of several researchers. Recall that the Theory of Island Biogeography as developed applied to continental islands. Our four theories have been applied to habitat islands or patches in an often not so benign matrix. These four theories lead to a General Theory of Insular Biogeograpy that makes a special case of the Theory of Island Biogeography. Edge, Juxtaposition, and Corridor Theories do not apply to islands; however, the External Impact Theory does apply. Many of the results of island biogeography apply to isolated continental fragments. However, whereas negative edge effects are now widely accepted as occurring in continental fragments, edge effects were not originally part of the Theory of Island Biogeography. We neither think of islands as being connected by corridors, nor juxtaposed with altered habitats. We should no longer rely on the crutch of the Theory of Island Biogeography to explain results that are only remotely similar to continental islands.

Application of the Theory

Assume there exists a metacommunity of species S^1 and species S^2 in five different landform cover types, C_1 to C_5. Generally, species use cover types differently.

We use the word habitat to refer to those cover types acceptable (in a broad sense) to a particular species. S^1 and S^2 utilize C_1 to C_5 differently according to Table 7.1. The collection of all cover types is referred to as the universe. Assume that a square or hexagonal grid overlays the universe and that each of 100 grid cells each contains a single cover type. Suppose that S^1 and S^2 occupy different amounts of each cover type and densities vary between these types according to Table 7.1. S^1 might be humans and S^2 wolves. Each perceives C_1 to C_5 differently.

Different species utilize cover types differently (see Table 7.1). Optimal habitat is prime habitat for a species. Suboptimal habitat is habitat that is less than optimal habitat, perhaps where reproductive and foraging success are high, but not optimal. Marginal habitat refers to habitat where the species can survive, but might not adequately reproduce. Invasible habitat is habitat not currently unoccupied, but could be if conditions change. Habitat that is not traversable acts as a barrier to dispersal and movement to the species and remains unoccupied. All but nontraversable habitat is assumed to be traversable, thus the number of traversable habitat cells is the sum of the number of optimal, suboptimal, marginal, and invasible cells.

TABLE 7.1

Cover type	Number of cells (N)	S^1 habitat type	Number per cell (2)	S^2 habitat type	Number per cell (2)
C_1	10	Optimal	20	Suboptimal	5
C_2	20	Suboptimal	10	Invasible	0
C_3	20	Marginal	5	Marginal	2
C_4	15	Invasible	0	Optimal	10
C_5	35	Nontraversable	0	Nontraversable	0

To compute average habitat quality for each species, habitats must have an associated value. We assume that each species values each cover type differently. First, we compute the *total population* of each species

$$T^1 = \sum_{i=1}^{5} N_i * S_i^1 = 10 * 20 + 20 * 10 + 20 * 5 + 15 * 0 + 35 * 0 = 500$$

$$T^2 = \sum_{i=1}^{5} N_i * S_i^2 = 10 * 5 + 20 * 0 + 20 * 2 + 15 * 10 + 5 * 0 = 240$$

The sums run across all cover types because the habitat for a particular species varies with the cover type. In general, the total population of S_j in i different habitats is given by:

$$T^j = \sum_{i=1}^{5} N_i * S_i^j$$

Average **habitat quality** over the region for S^j can be calculated by assigning values to each habitat. Let optimal habitat have a value of 8, suboptimal habitat a value of 6, marginal a value of 4, and invasible 2. Nontraversable habitat has a value of 0. Note that C_1 above is optimal habitat for S^1 and so has a value of 8 while simultaneously has a value of 6 for S^2 because the habitat is suboptimal for S^2. Let $v_{i,j}$ be the weighting assigned to habitat i for S^j.

$$Q^j = (1 / 100) * \sum_{i=1}^{5} V_{i,j} * N_i$$

For instance,

$$Q^1 = (1 / 100) * [(8 * 10) + (6 * 20) + (4 * 20) + (2 * 15) + (0 * 35)]$$

$$= (1 / 100) * 310 = 3.1$$

For S^2,

$$Q^2 = (1 / 100) * [(6 * 10)] + (2 * 20) + (4 * 20) + (8 * 15) + (0 * 35) = 3.0$$

Overall, the area occupied by S^2 is of lower quality because of the large number of suboptimal habitat cells. Habitat quality can be weighted by the population residing in the habitat:

$$Q^2 = (1 / T^j) * \sum_{i=1}^{5} V_{i,j} * N_i * S_i^j$$

We find

$$Q^1 = (1 / 500) * [(8 * 10 * 20) + (6 * 20 * 10) + (4 * 20 * 5) + (2 * 15 * 0) + (0 * 35 * 0)] = 6.4$$

and

$$Q^2 = (1 / 240) * [(6 * 10 * 5) + (2 * 20 * 0) + (4 * 20 * 2) + (8 * 15 * 10) + (0 * 35 * 0)] = 6.9$$

$Q^2 > Q^1$ because a higher percentage of the total population of S^2 occupies higher quality habitat than does the total population of S^1.

Habitat connectivity can be measured as the fraction of the universe occupied by traversable cells. If the grid is regular (rectangular, hexagonal) we can then assign a probability that a corridor exists through the universe using the results from the Percolation Theory. Note that habitat connectivity depends not on cover type, but on the habitat type and is thus dependent on the particular species. For S^j, habitat connectivity, HC, is:

HC^j = (number of cells in universe -)/(number of cells in universe)

where the ith cover type is nontraversable habitat. Hence

$$HC^1 = \left(100 - C_i^1\right)/100 = (65/100) = 0.65$$

and

$$HC^2 = \left(100 - HC^1\right)/100 = 0.45$$

Results from the Percolation Theory suggest that S^1 will be able to traverse the universe while S^2 will not find a suitable corridor that spans the universe.

Habitat fragmentation is defined as the fraction of the universe that is nontraversable habitat:

$$HF^i = H_i^1 / \text{(numbers of cells in the universe)}$$

where the ith cover type is nontraversable habitat.

Note that habitat fragmentation when added to habitat connectivity sums to unity:

$$HF^i + HC^i = 1$$

Often a landscape appears to have suitable cover types, but the organism of particular interest is not present. Although trite, things are not always what they appear to be. We can slice, dice, and categorize landscape features and cover types (Gustafson 1998). However, we prefer to provide an example of landscape contextual analysis. Figure 7.1 shows a hypothetical landscape overlaid with 100 hexagonal cells. Each grid cell is assigned a habitat value for a particular organism. At first appearance, the landscape appears to have many favorable cells, and one might conclude that populations of the particular organism of interest would be healthy. The classification is similar to that used above; however, we have adapted it for a contextual analysis as follows.

Our contextual analysis will be based on a set of rules depending on the "sphere of influence" that different cover types have on a particular organism. The organism-specific rules will be applied in order. For the hypothetical organism used here, detrimental cells have a sphere of influence greater than the space they occupy. For example, sound from these detrimental cells might

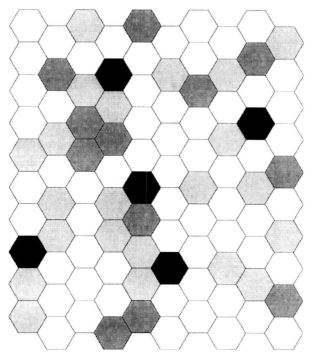

FIGURE 7.1
A fragmented landscape of 100 hexagonal cells. Empty cells are optimal habitat, light gray are suboptimal, darker gray are marginal, and black are detrimental.

Color	Cover type	% of landscape
White	Optimal	61
Light gray	Suboptimal	22
Dark gray	Marginal	12
Black	Detrimental	5

travel across the landscape and impact the particular organism negatively. For other organisms, this sound might have no influence and so the sphere of influence of the detrimental cells would be less. To account for this influence, all neighboring cells will be changed to marginal from whatever classification they were assigned.

> **Rule 1** (Juxtaposition Theory): All neighboring cells of detrimental cells will be assigned as marginal. Marginal habitat also has a sphere of influence beyond its border.

> **Rule 2** (Juxtaposition Theory): All neighboring cells of marginal habitat will be assigned suboptimal. Thus, detrimental cells affect not

only their immediate neighbors, but also their once-removed neighbors.

Rule 3 (Juxtaposition Theory): Marginal cells reduce their optimal neighbors to suboptimal. Suboptimal cells have edge effects that are damaging to optimal cells.

Rule 4 (Edge Theory): Suboptimal cells create edge effects in neighboring optimal cells.

The result of applying a contextual analysis to the hypothetical landscape in Figure 7.1 yields Figure 7.2 with:

Color	Cover type	% of landscape
White	Optimal	5
Light gray	Suboptimal	51
Dark gray	Marginal	39
Black	Detrimental	5

Although the landscape in Figure 7.1 appeared to have many optimal and suboptimal cells, the contextual influence of marginal and detrimental habitat and edges effects considerably reduced the number of these habitats.

Furthermore, the influence of detrimental cover types often extends differentially in one or more directions, or can leapfrog across a landscape such as happens when fire in sugar cane fields carries nutrients deep into the southern Everglades. In this case, the influence of detrimental cells extends 100 km or more during particular seasons. Negative edge effects also reduce available favorable habitat (Figure 7.2).

Obviously, more complex rules can be applied to the contextual analysis of landscapes. These rules can be empirically derived in some cases. Contextual analysis enables an analytic exploration of landscapes beyond content and appearance. In the case of the Florida Everglades, detrimental areas surrounding the national park have a large sphere of influence that can now be quantitatively studied. Contextual analysis can be applied to study the migration route of the monarch butterfly, for instance, because we can extend the analysis of content across the landscape based upon a set of rules derived from theory that are species specific.

Habitat Conservation Plan

Section 10 of the Endangered Species Act of 1973 was amended in 1981 to include that each designation of a threatened or endangered species required the creation of a habitat conservation plan (HCP). An HCP is a written document that specifies how much land must be set aside to protect threatened

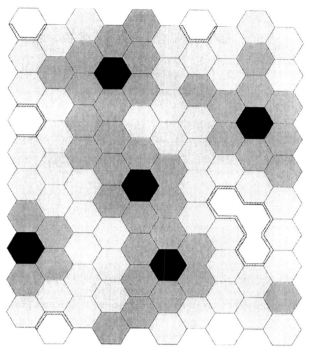

FIGURE 7.2

A fragmented landscape of 100 hexagonal cells after contextual habitat analysis. Empty cells are optimal cells, light gray are suboptimal, darker gray are marginal, and black are detrimental. Edge effects are also shown.

and endangered (commonly abbreviated as T&E) species. A recent example served to illustrate how HCPs were presumed to work. The Alabama beach mouse was listed as federally endangered in 1985 (Kaiser 1997). The U.S. Fish and Wildlife Service published an HCP within two years to protect the 142 ha. of beach where the mice made their last stand. Later a developer petitioned the Service for permission to construct a resort community on part of the mouse's protected beach. In exchange, some dunes would be restored and signs would be put up reminding beachgoers that an endangered mouse lived nearby. In addition, a fund was to be established to support research on the mice.

In April 1997, Reed Noss of Oregon State University and a group of prominent conservation biologists stated in a letter to environmental leaders in Congress and the White House that many of the HCPs had been developed in the absence of sound scientific input from scientists. Furthermore, of the concept of HCPs the biologists stated "They're not only appropriate, but the only way to go." One month later a U.S. Geological Survey biologist filed a lawsuit against the U.S. Fish and Wildlife Service challenging the new resort development on the grounds that the remaining beach and dunes would not support a viable population of mice (Shilling 1997). Can an isolated fragment save the Alabama beach mouse?

HCPs attempt to legitimize setting aside small fragments as if they might protect something within them in perpetuity. From a landscape ecological point of view, we know this cannot work. Habitat fragments lay in a background landscape and therefore are connected to and influenced by nearby landscapes and the activities that occur within them. Other processes such a hurricanes, excessive winds, oil spills, and stochastic and unforeseen events happen. The probability of these events happening sometime in the future is unity. The probability that some event will have a disastrous effect on the Alabama beach mice increases as the size of the fragment they reside in decreases and increases as the distance to human development decreases.

We can apply Juxtaposition Theory and the External Impact Theory and make a prediction of the future of proposed reserves. We can say with certainty that one of two outcomes will occur. External Impact Theory implies a natural episodic event, probably a hurricane, will lay waste to what remains of the 142 ha. Alabama beach mouse habitat, HCP notwithstanding; Edge Theory and Juxtaposition Theory suggest that human activities in and around the resort causing excessive pollution in the form of noise, light, pesticide, and lawn care chemicals, or simply people strolling on the beach will eventually push the mice to extinction. More than likely, a loose pet cat will ignore the signs and do the mice in. In other words, no amount of "good" or even "great" science will serve to protect some pitiful fragment of what was once continuous Gulf beach from Texas to the Florida Keys. The fact of the matter is that many beachfront specialists, among them the Alabama and Florida beach mice, and several seaside sparrow subspecies such as the Cape Sable seaside sparrow are in trouble or already extinct as is the Dusky seaside sparrow. Are pupils of landscape ecology supposed to learn four theories that exist on paper and then ignore the results of the theories in practice?

Re-Membering Fragmented Landscapes

Harris and Scheck (1991) wrote that conserving isolated natural areas in lieu of interconnected landscape systems was doomed to failure. Shafer (1994) and others agreed. Creating movement corridors in human-dominated landscapes is one way of protecting natural ecological functions (Harris and Scheck 1991; Zonneveld 1994). Huffaker (1958) demonstrated experimentally that corridors worked to prolong predator–prey interaction. Preston (1960), MacArthur and Wilson (1963, 1967), and Diamond (1975) developed theoretical bases for movement corridors. Merriam (1984) showed that populations persisted longer in interconnected landscapes. Decades of empirical evidence led ecologists to conclude that tracts of nature that are physically interconnected to larger source pools of organisms support and maintain a greater richness of organisms than comparable tracts that are not physically connected.

Evidence now suggests that conservation of native fauna and flora will be achieved when natural levels of gene flow between organisms and populations are maintained across landscapes. Schemes that either artificially restrict or increase gene flow (such as translocation programs) above natural levels will not work (Frankel and Soulé 1981; Schonewald–Cox et al. 1983; Oldfield 1989).

Restoring Landscape Processes: The Case for Movement Corridors

We believe the case for movement corridors is by now well established (Forman 1995; Hudson 1991; Harris and Gallagher 1989; Harris 1988). Current efforts should be made to integrate large areas by creating movement corridors connecting these large areas. By focusing on the landscape, we are in fact focusing on preserving ecological processes and placing less emphasis on saving species. We are forced to consider community patches in a spatially explicit way. Use of the migration corridor approach was critical to restoration of the waterfowl populations of North America. Fish ladders were designed to facilitate movement.

The bottom-up effects of landscape fragmentation such as loss of favorable habitat, increased isolation, increased negative edge effects, and the effects of introduced species have been disastrous. The negative effects of increased isolation of grassland fragments demonstrated that important ecological processes could be disrupted. Though prairies were once common in Wisconsin, they have been reduced to dismembered fragments today. Extinction rates of plants in these remaining prairie fragments was 0.5 to 1% per year (Leach and Givnish 1996). Though the Theory of Island Biogeography could again be applied to explain the loss of plant species, the importance of terrestrial connectivity must be appreciated. Previously, wildfires were common on the prairies. As prairie fragments became more isolated, wildfires occurred less frequently and on much smaller areas. Moreover, nonflammable barriers such as roads, agricultural fields, shopping centers, and housing developments prevented fires from spreading.

Fire suppression and the complete loss of naturally spreading fires in fragmented prairies increased local loss of short species, and tall and woody species numbers increased. Species with smaller seeds showed higher rates of extirpation than those with larger seeds. Herb species declined in both dry and wet prairies. The loss of the ecological process of fire due to fragmentation changed the prairie plant community and may well be negatively impacting other dismembered landscapes elsewhere (Leach and Givnish 1996).

Natural recolonization and recovery programs have led to an increase in wolf populations in Minnesota, Wisconsin, and Michigan. In the late 1970s biologists observed that the growing Minnesota wolf population had begun to disperse, recolonizing northern Wisconsin, and, later, upper Michigan from Wisconsin and Canada (Hammill 1995; Mech and Nowak 1981). Some natural recolonization had occurred in northwest Montana (Ream et al. 1991). Successful wolf recovery in the Lake States region was largely a result of legal protection and changed public attitudes of greater tolerance and acceptance of wolves' existence in the wild (Mech 1995). Although wolf recovery in the Lake States region is a positive accomplishment, restoration of the top carnivore is not a sign that the former forest ecosystem is also restored to some previous level of functioning.

To better plan for and manage the future wolf population, the likely areal extent and magnitude of regional wolf recovery needs to be understood in a more spatially explicit manner than has been explored thus far. This involves analyzing landscape-scale factors important to the suitability of potential wolf range and future wolf population recovery levels and assessing possible interactions of wolves with people and other aspects of the regional landscape using contextual analysis techniques described earlier.

Another human-caused effect, this one indirect, was the importance of high ungulate prey populations, usually white-tailed deer (*Odocoileus virginianus*) in the Lake States, to successful wolf recolonization (Fuller 1995). Ungulate levels were particularly high in the Lake States because deer thrived in the altered and fragmented habitat—young, managed forests with ample openings—of northern Wisconsin and upper Michigan (Fuller 1995; Fuller et al. 1992; Mladenoff and Sterns 1993). However, recent work (deCalestra 1994, 1995) has shown that the high deer population that benefits wolves can negatively affect other important aspects of forest biodiversity and ecosystem functions (Smithsonian Institution 1994). Deer levels were too high and negatively affecting other forest species (Vander Zouen and Warnke 1995). High levels of deer browsing had direct impacts on palatable species, such as understory plants (Balgooyan and Waller 1995; deCalestra 1995).

High levels of deer browsing favoring certain tree species over others altered forest regeneration and composition. In the Lake States, these changes caused ecosystem feedbacks through altered forest floor litter composition and quantity. These alterations caused changes in nutrient cycling dynamics (cite). Severe understory browsing also negatively affected forest habitat structure for birds (deCalestra 1995). For instance, reductions in insectivorous forest birds resulted in greater levels of insect herbivory on forest trees, reducing tree growth and productivity (Marquis and Whalen 1994). In developed, semiagricultural landscapes, forest fragmentation results in increased bird predation and nest parasitism associated with forest habitat edges (Brittingham and Temple 1983; Wilcove 1985). Because deer abundance was also favored in these fragmented, mixed landscapes, deer browsing at high levels contributed significantly to forest bird impacts in these regions. The detrimental effects on other species of local deer abundance were documented in many areas (Anderson and Katz 1994).

This growing understanding of the complexity of ecosystems suggests that they must be managed in ways that better integrate commodity production and human needs with concern for long-term sustainability and ecosystem functioning, of which biodiversity protection is an integral part (Franklin 1993; Mladenoff et al. 1994). Clearly, conservation decisions must also be placed in a larger landscape context.

Are parks and protected areas the future of biodiversity preservation? By now the answer should be clear. Newmark (1987) made the case that many of the largest parks of North America are simply too small. Shafer (1994) summarized some of the relevant issues. In most cases, if the goal is to restore biological processes over large areas, then wildlife must be protected outside parks and protected areas. If we value the process of evolution, then we must learn to live with biodiversity, all biodiversity, beyond protected areas.

Wolf Reintroduction

Recent research provides evidence that wolves are functionally important in northern forest ecosystems by exerting top-down control in forest food chains (McLaren and Peterson 1994), emphasizing the importance of maintaining and restoring formerly abundant species that are now endangered. The wolf is illustrative of a wide-ranging, potentially abundant species that must be managed as a part of the greater semiwild managed forest matrix, and not in small reserves. Although reserves remain essential and useful for many conservation objectives (Noss 1993), they are inadequate for many large-scale needs.

Landscape-based conservation perspectives have implications for the restoration of top predators. The spatial distribution of favorable habitat has been mapped with a geographic information system (Mladenoff et al. 1997, p. 23). It is clear that these habitats are not connected. Wolf packs avoided certain land cover types, such as agriculture and deciduous forest, and favored forests with at least some conifer component, such as mixed deciduous–conifer forest areas and forested wetlands. Public lands, particularly country forests, were preferred, whereas private lands were avoided. Centers of wolf pack territories were most likely to occur in areas with road densities below 0.23 km/km^2, and nearly all wolves occurred where road densities were less than 0.45 km/km^2. No wolf pack territory was bisected by a major highway. Human population density was less than 1.52 individuals per square kilometer in the areas favored by wolves (Mladenoff et al. 1995).

Two methods were used to estimate potential wolf populations: estimating the overall wolf population by relating total potential habitat area to mean pack territory size, and spatially estimating potential wolf density by considering wolf population and prey density relationships (Fuller 1995, Fuller et al. 1992). Can estimate spatial distribution and abundance.

Wolves and other top predators play an important role in natural ecosystems (Peterson 1988), and wolf recovery in the Lake States is a positive accomplishment that has been managed under the Endangered Species Act. Wolf recovery is a conservation success built on both direct and indirect human influences, and it has an impact on human society and other aspects of forest biodiversity besides wolves. Consequently, we need to consider such conservation efforts in a landscape context because of the following implications.

Wolves Require Management

The ecology and large range of wolves dictate that recovery of sizable populations must take place not in small, isolated reserves, but in the large matrix of managed, human-dominated lands. Therefore, wolf recovery is particularly dependent on human attitudes. If wolves are not killed, and ungulate prey are adequate, they can apparently occupy semiwild lands formerly thought to be unsuitable (Fuller 1995; Mech 1995). The dispersal ability and adaptability of wolves will allow them to colonize increasingly developed areas (Mech 1995; Mech et al. 1995).

A second conservation conundrum relates to the ways that current landscapes now being recolonized by wolves are different from the original presettlement conditions where wolves, and other carnivores, were previously widespread. Deer occupy human-disturbed landscapes and are the prey base that attracts wolves. This brings wolves into contact with humans.

There is a strong inverse relationship between the prey (ungulate) population size and the size of a wolf pack territory. Where prey levels are higher, a given pack requires a smaller area to meet its needs (Fuller 1995; Wydeven et al. 1995).

Colonization of the fragmented habitat in Wisconsin may remain dependent on source–sink dynamics (Pulliam 1988) with the saturated Minnesota population. With legal protection, high rates of dispersal from the high Minnesota population have maintained continued colonization of the less favorable habitat in Wisconsin with its higher mortality.

Recent evidence shows that wolves are moving long distances (hundreds of kilometers) from within far northern Minnesota to as far as south-central Wisconsin and upper Michigan, across large areas of habitat unsuitable for colonization (Mech 1995; Mech et al. 1995). Simple binary models of landscapes with suitable habitat islands and corridors within an unfavorable matrix (Simberloff et al. 1992) may apply poorly to the wolf, a wide-ranging top carnivore that is not habitat specific. Landscapes may be better viewed as a probability surface of varying suitability through which animals move and colonize. Continued recovery of wolf populations in Wisconsin depend on wolves dispersing long distances from Minnesota. We must understand the complex interactions and trade-offs that may be required to balance biodiversity conservation with continued demand for resources from our landscapes.

The reintroduction of the wolf in the Lake States demonstrated that humans and wolfs can coexist across the landscape. Wolfs were not restricted to isolated parks or reserves, but were able and willing to travel long distance to recolonize favorable areas. If left alone, wolves will act to re-member a landscape mosaic and reestablish the ecological processes that maintain a healthy environment. This is an example of a landscape ecological approach to conservation. Other less acceptable approaches to conservation also exist.

Migration Corridor Identification

Dispersal corridors are "an essentially continuous band or congenial habitat by which many ecologically compatible species might extend their ranges" (Stehli and Webb 1985:3). It is little wonder that development of migratory corridor planning and management, and the system of stepping-stone wildlife refuges became the most critical factors in the restoration of migratory waterfowl populations in North America during the early decades of this century. As early as 1940, ecologists recognized that land bridges that physically interconnected otherwise disjunct faunal communities were critical to the range expansion of terrestrial fauna from one region to another (Simpson 1940). Connections consisting of stepping stones only (e.g., island chains) and land connections of short duration or other idiosyncratic features (e.g., a high elevation or high latitude) were observed to "filter" the fauna and result in unbalanced or highly biased subgroups of species. How extensive must connectivity be?

The migration of the monarch butterfly (*Danaus plexippus*) has become an endangered biological phenomenon (Brower 1995). Hamilton (1885) reported butterfly accumulations along the New Jersey coast:

> almost past belief...millions is but feebly expressive...miles of them is no exaggeration.

Shannon's (1916) description was also typical of early observations:

> ...wide highways of the Great Plains and West Central States offer the most frequent reports of remarkable butterfly spectacles...gatherings of almost unbelievable magnitude...move forward in congregations miles in width forming veritable crimson clouds.

In the west, Orr (1970) commented:

In Washington in 1928 a flock of monarchs estimated to be several miles wide and ten to fifteen miles long was observed in the Cascade Mountains. The number of individuals in this flock was believed to be in the billions.

Brower (1995) claimed that the monarch migration expanded eastward during the late 19th century. In 1837 the John Deere steel plow was introduced and by the 1880s the 20-mule combine harvester made possible large-scale farming. Plowing destroyed 433 million acres of the original midwestern prairie that was host to 22 habitat-specific and nonweedy *Asclepias* species (milkweed species for which the monarchs were adapted). While the prairie was being destroyed, the eastern forests were being replaced by open farmland that was host to a single milkweed species, *A. syriaca*, allowing the monarchs to shift eastward to a growing food source.

Two migratory populations of monarch butterflies occur in North America. Western populations have a summer range west of the Rocky Mountains and north to the Canadian border. These butterflies winter along the California coast. The much larger eastern populations summer east of the Rocky Mountains to the Canadian border and overwinter in 12 high fir forests in the Transverse Neovolcanic Belt of central Mexico.

Spring remigrations of the monarch butterfly occur in both populations. The butterflies lay their eggs on the resurgent milkweed and produce a spring generation. The new generation flies northward toward Canada to feed on emerging milkweed, laying eggs along the migration route. After the first spring and two or three subsequent summer generations the monarchs begin their southward migration to their overwintering grounds.

Monarch butterfly protection is a daunting task that must take place outside national parks and reserves and across the international landscape (Harris 1993). Clearly, overwintering sites (hot spots) require protection. But is this enough? Monarchs are milkweed specialists and these plants are actively destroyed by farmers. While the Great Plains host 20 species of *Asclepias*, the eastern fields have but one. Eastern forests are known to be regenerating, reducing habitat for milkweeds and their hosts. Can a series of so-called stepping-stone reserves be established across the Great Plains and west of the Rocky Mountains? What about the numerous hazards monarchs face on their travels? Highways criss-cross much of their habitat, making migrations a risky business indeed. "It is now abundantly clear that if we do not meet the challenge of landscape ecology, we can harbor little hope of stanching anthropogenic losses in the Earth's biodiversity and hence of stemming the deterioration in the life support system of our own species" (Lidicker 1995).

Simberloff (1998) suggested that ecologists identify keystone species and attempt to elucidate the mechanisms that cause them to contribute disproportionately to ecosystem functions. Gaining an understanding of the ecology of landscapes requires consideration of mobile species. Highly mobile species are sometimes carnivores that require large natural or seminatural areas to survive. These species might use corridors and so would benefit from re-membering landscapes. We do not believe that the full complexity of nature will be revealed anytime soon; however, we must not delay our quest to appreciate these complexities and put what we learn into practice.

We disagree with Simberloff (1998) that many threatened species, including flagship species like the spotted owl, the red-cockaded woodpecker, and the Florida panther could disappear entirely from an ecosystem without major or even detectable changes in key processes. We suspect our differences, however, have to do with time spans. Certainly, we agree with Simberloff that, over the short term, the disappearance of spotted owls from forests would go unnoticed. However, we subscribe to the belief that over ecological time the disappearance of owls would lead to changes in ecosystem dynamics, most likely indirectly through their prey species, and so produce noticeable, quantifiable changes to the forest. The short lifespans and current state of ignorance of humans, however, do not allow us to support our belief without referring to the biological evolution. We simply do not believe that evolution produces superfluous organisms, and that is a compelling reason for urging a top-down approach to landscape ecology as Simberloff advocates.

Putting Things Together: An Ecology of Landscapes

To anchor the study of the ecology of landscapes evolution must play a crucial role. The evolution of the horse is an example of how species evolve across space and through time. The races of red-shouldered hawks demonstrate an example of a clinal species, that is, a species that shows morphological changes over space at a particular time. Different numbers of species coexist across space and through time. For a single snapshot in time we might see a biodiversity hot spot (Figure 7.3). Note that the x-axis represents space, but that it could represent time as well.

Graham et al. (1996) showed that species move through time somewhat independently in a Glassiness way. However, local communities persist over ecological time and multilevel selection (Wilson 1997) acts on individuals within communities, leading to a more Commendation view of locally adapted organisms. Multilevel selection allows species in communities to develop unique traits when characters associated with those traits are genetically pliable. Rudely (1993, p. 234) compared piable morphological and behavioral traits. Though selection acts on individuals, selective forces are the integration of all such forces and are played out locally.

For example, Geoffroy's cats (*Oncifelis geoffroyi*) weigh 2.5 kg in Paraguay and 5 kg in Patagonia. The crested caracara (*Caracara plancus*) of Patagonia is half again as large as the Florida crested caracara. The plumage of both is nearly identical, but behavioral differences exist. The Pampas cat (*O. colocolo*) of South America has many very different coat patterns, from one with no spots to one nearly covered in spots. Given a particular coat pattern the geo-

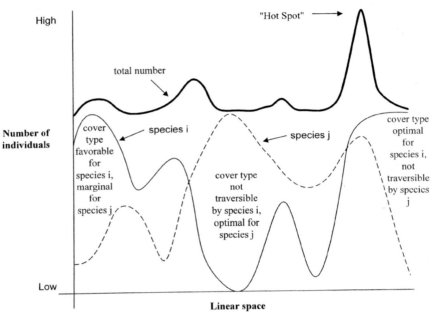

FIGURE 7.3
Quality of space (here one-dimensional) and other variables determine the local abundance of different species. Quality of the same space varies between species. Here species i (solid line) and species j (dashed line) abundances are shown. Total abundance (bold line) can be used to identify "hot spots" in space. Note that the x-axis could refer equally as well to time instead of space.

graphic location of the cat can be determined. This is because evolution is acting locally on pliable characters in a unique local community.

Keystone species (Simberloff 1998) and ecosystem engineers (Jones et al. 1994) are examples of organisms whose behavior or activities affect ecosystem functions in a disportionate way. Top carnivores often act to organize prey populations in space and so also affect local resources disproportionately. Prey populations such as deer eat vegetation and can in turn affect the distribution and abundance of insects and hence insectivorous birds.

Willson et al. (1998) suggested that anadromous and inshore-spawning marine fish that provided a rich, seasonal food resource affected the biology of both aquatic and terrestrial consumers and indirectly affected the entire food web that tied water to land. Willson et al. referred to the fish as a "cornerstone species" because of the disproportionate resource they provided to the coastal water–land ecotone. Marine-derived nutrients passed from the fish to birds of prey, terrestrial carnivores such as bears, into the soil via invertebrates, and then into plants. With the large reduction of many fish species due to anthropocentric activities, top predators disappeared and reductions in vegetation occurred, in turn providing less insect prey for young fish.

Each of the above examples goes well beyond the formation of landscape patterns and the effects of patterns on ecosystem processes (what we have referred to as landscape effects). In each of our examples evolution, most likely multilevel species evolution, was at work in multiple ecosystems and hence on the landscape. We can generalize these examples into the following theories that allow hypotheses to be generated and tested.

Example

Consider a forest adjacent to an agricultural field. The agricultural field is supported and maintained by fertilizer, pesticide, and large quantities of water. After harvest the field is abandoned. The forest begins to encroach onto the field using an arsenal of weapons. Tree roots beneath the ground sequester moisture, delivering it to the growing trees. Some trees might shade the field, preventing energy from the sun from reaching the new plants sprouting in the field. Mobile organisms such as birds hawk insects over the field and deposit the waste products in the forest, while other birds drop shrub seeds far out into the field. A small forest cat might kill chickens feeding in the field, carrying its prey back into the forest to consume. In this way the forest competes successfully against field. The organisms of the forest, acting as individuals going about their daily lives, war together against the field. As trees conquer the field, their fitness increases because they leave more offspring, as do the birds that use the forest. The forest community can be thought of as an entity competing against another entity. From this view, we can see where each component of the entity fits. We can hypothesize that were we to remove the forest cat, the forest would be less successful against the field. We might propose an experiment that, over the course of 100 years, could test our hypothesis. From our viewpoint we can see more clearly how organisms are acting in the forest, not just as stagehands, but as mobile agents of the forest ensemble, capable of extending the influence of the whole, acting as individuals to increase their own fitness and not under the guidance of an unknown coordinating force. Some might consider this to be self-organized complexity, dynamic equilibria, or a complex adaptive system. Certainly, complex interactions are an important source of variation on which selection can operate at the level of the local community. In our view each species acts in a Gleasonian fashion, creating an ensemble that appears to function in a Clementsian fashion. We believe that one need look no further than evolution to resolve these research questions.

We conclude by stating three additional hypotheses:

> **Landscape Biodiversity Hypothesis**—Multilevel species adaptations in local communities leads to unique gene pools across the landscape. Community gene pools differ according to the degree of landscape connectivity (a species-dependent parameter), the separation time, and quantifiable external factors.

Ecosystem Organism Hypothesis—All organisms exert top-down effects on ecosystem processes. The removal or replacement of any organism produces a measurable response in ecosystem variables.

Landscape Organism Hypothesis—Mobile organisms exert top-down influences in all utilized ecosystems.

We have given specific theories and hypotheses that we hope will organize a research program in landscape ecology that extends beyond the matrix-patch-corridor landscape ecology paradigm of the past (Forman and Godron 1986). We did not specifically discuss regional ecology and we do not suggest that scaling up landscapes produces biomes. Just as no one has suggested scaling an individual creates a population, scaling an ecosystem does not produce a landscape.

8

Quantifying Constraints upon Trophic and Migratory Transfers in Landscapes

CONTENTS

Ecosystems are neither machines nor superorganisms, but rather open systems that require a "calculus of conditional probabilities" to quantify. Autocatalysis, or indirect mutualism, as it occurs in causally open systems, may act as a nonmechanical, formal agency (sensu Aristotle) that imparts organization to systems of trophic exchanges. The constraints that autocatalysis exerts upon trophic flows can be quantified using information theory via a system-level index called the ascendency. This quantity also gauges the organizational status of the ecological community. In addition, the ascendency can be readily adapted to quantify the patterns of physical movements of biota across a landscape. In particular, one can use ascendency to evaluate the effects of constraints to migration, even when the details of such constraints remain unknown.

Introduction

In his recent critique of ecology, Peters (1991) warns ecologists to pursue only those concepts that are fully operational. In a strict sense, a concept is fully operational only when a well-defined protocol exists for making a series of

1-56670-368-9/00/$0.00+$.50

measurements that culminate in the assignment of a number, or suite of numbers, that quantifies the major elements of the idea. Can the ascendency description of ecosystem development be applied to spatial heterogeneities in ecosystems in a way that will yield fruitful insights and/or predictions?

In a recent book (Ulanowicz 1997) I attempted to articulate the full meaning, import, and application of "ecosystem ascendency" as a quantitative description of development in ecosystems. But the section in that volume that dealt with spatial heterogeneities is notable for its brevity and dearth of specific examples. Whence the attempt through what follows to elaborate more fully the potential for employing information theory in landscape ecology. Before proceeding with quantitative definitions, however, it would be helpful to review briefly the conceptual background into which any theory of ecosystems must fit.

Conceptual Background for Ecosystems

According to Hagen (1992), three metaphors have dominated the description of ecosystems (Figure 8.1): (1) the ecosystem as machine (Clarke 1954; Connell and Slatyer 1977; Odum 1971); (2) as organism (Clements 1916; Shelford 1939; Hutchinson, 1948; Odum 1969); and (3) as stochastic assembly (Gleason 1917; Engelberg and Boyarsky 1979; Simberloff 1980). Hagen portrays the debates among the schools that champion each analogy in terms of a three-way dialectic—an antagonistic win/lose situation. He sees, for example, the

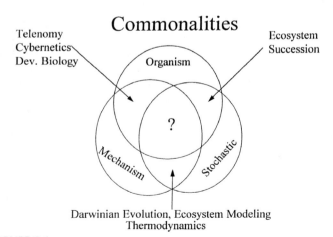

FIGURE 8.1
A Venn diagram depicting overlaps among the three major metaphors for ecosystems. (After Hagen 1992. With permission.)

holistic vision of Hutchinson and E.P. Odum as having been gradually displaced during the 1950s and 1960s by the disciples of the neo-Darwinian/nominalist synthesis.

By way of contrast, Golley (1993) believes that holism in ecology is alive and well. According to Depew and Weber (1994), for example, Clements inadvertently provided the nominalists with lethal ammunition by casting the ecosystem as a "superorganism." Apparently, Clements conflicted physical size and extent with organizational complexity in drawing his unfortunate analogy. If, however, one reverses Clements' phraseology and instead characterizes "organisms as superecosystems," then much of the criticism against holism in ecology is circumvented.

It is pressing the ecosystem metaphors beyond their intended limits that causes many to regard these images as mutually exclusive, and to conclude that truth can lie in only one corner of the triangle, none of which is to suggest that reality (insofar as we are capable of perceiving it) occupies the middle ground. Rather it is to perceive nature as being somewhat more complicated than has heretofore been assumed, and to propose that any adequate description of development in living systems must be *overarching* with respect to simplistic analogs.

As a first step towards amalgamating these analogies, it is useful to consider the commonalities and differences among the metaphors. Of the three, the one most familiar to readers is bound to be the mechanical, for it is the analogy that has driven most of modern science. Depew and Weber (1994) (see Table 8.1) cite four assumptions that undergird the Newtonian worldview: (1) the domain of causes for natural phenomena is *closed*. More specifically, only material and mechanical causes are legitimate in scientific discourse. (2) Newtonian systems are *atomistic*. That is, they can be separated into parts; the parts can be studied in isolation; and the descriptions of the parts may be recombined to yield the behavior of the ensemble. (3) The laws of nature are *reversible*. Substituting the negative of time for time itself leaves any Newtonian law unchanged. (For example, a motion picture of any Newtonian event, when run backwards, cannot be distinguished from the event itself.) (4) Events in the natural world are inherently deterministic. So long as one is able to describe the state of a system with sufficient precision, the laws of nature allow one to predict the state of the system into the future with arbitrary accuracy. Any failure to predict must result from a lack of knowledge.

To Depew and Weber's four pillars of Newtonianism one must add a fifth assumption, *universality* (Ulanowicz 1997). Newtonian laws are considered valid at all scales of space and time. Whence, physicists have no qualms (as perhaps they should) about mixing quantum phenomena with gravitation (Hawking 1988).

When one regards the nominalists' presuppositions, we find them more simple still. Stochasticists agree with Newtonian that causality is closed (only material and mechanical forms allowed) and that systems are atomistic (virtually by definition). But they regard the remaining three assumptions as

TABLE 8.1

Comparisions of Outlooks

Mechanism (*Newtonianism*)	Organism (*Holism*)	Stochasticism (*Nominalism*)
Material, Mechanical	Material, Mechanical	Material, Mechanical
	Formal, Final	
Atomistic	Integral	Atomistic
Reversible	Irreversible	Irreversible
Deterministic	Plastic	Indeterminate
Universal	Hierachial	Local

unnecessarily restrictive and so consider events to be irreversible, indeterminate, and local in nature.

The organismal or holistic worldview differs most from the other two and requires elaboration. Critics of holism, of course, will immediately invoke Occam's Razor as they inveigh against what they regard as wholly unnecessary (and, in their own eyes, illegitimate) introductions. One must bear in mind, however, that Occam's Razor is a double-edged blade, and that those too zealous in its application always run the risk of committing a Type-2 error by excising some wholly natural elements from their narratives.

Unlike the second Newtonian axiom, organic systems (again, almost by definition) are not atomistic, but integral. One cannot break organic systems apart and achieve full knowledge of the operation of the ensemble operation by observing its parts in isolation. Common experience provides no reason why organic systems should be considered reversible. As regards determinacy, in this instance the organic view does lie midway between the other two. The prevailing holistic attitude would probably describe organic systems as "plastic." One may foretell their form and behavior up to a point, but there exist considerable variations among individual instantiations of any type of system or phenomenon. This degree of "plasticity" may vary according to type of system. For example, the Clementsian description of ecosystems as superorganisms implied a strong degree of mechanistic determinism, whereas Lovelock's (1979) description of how the global biome regulates physical conditions on earth appears quite historical by comparison.

But what of causal closure? If causes other than mechanical or material may be considered, does this not automatically characterize the organic description as vitalistic or transcendental? Certainly, to introduce the transcendental into scientific discourse would be to defy convention, but it will suffice simply to point out that the idea of closure is decidedly a modern one. Aristotle, for example, proposed an image of causality more complicated than the current restricted notions. He taught that a cause could take any of four essential forms: (1) material, (2) efficient or mechanical, (3) formal, and (4) final. Any event in nature could have as its causes one or more of the four types. One example is that of a military battle. The material causes of a battle are the weapons and ordnance that individual soldiers use against their enemies. Those soldiers, in turn, are the efficient causes, as it is they who actually swing the sword, or pull the trigger to inflict unspeakable harm upon each other. In the end, the armies were set against each other for reasons that were economic, social, and/or political in nature. Together they provide the final cause or ultimate context in which the battle is waged. It is the officers who are directing the battle who concern themselves with the formal elements, such as the juxtaposition of their armies via-a-vis the enemy in the context of the physical landscape. It is these latter forms that impart shape to the battle.

The example of a battle also serves to highlight the hierarchical nature of Aristotelean causality. All considerations of political or military rank aside, soldiers, officer, and heads of state all participate in the battle at different scales. It is the officer whose scale of involvement is most commensurate with those of the battle itself. In comparison, the individual soldier usually affects only a subfield of the overall action, whereas the head of state influences events that extend well beyond the time and place of battle. It is the formal cause that acts most frequently at the "focal" level of observation. Efficient causes tend to exert their influence over only a small subfield, although their effects can be propagated up the scale of action, while the entire scenario transpires under constraints set by the final agents. Thus, three contiguous levels of observation constitute a fundamental triad of causality, all three elements of which should be apparent to the observer of any physical event (Salthe 1993). It is normally (but not universally, e.g., Allen and Starr 1982) assumed that events at any hierarchical level are contingent upon (but not necessarily determined by) material elements at lower levels.

One casualty of a hierarchical view on nature is the notion of universality. The belief that models are to be applicable at all scales seems peculiar to physics. If a physicist's model should exhibit a singularity whereby a phenomenon of cosmological proportions, such as a black hole, might exist at an infinitesimal point in space, then everyone soberly entertains such a possibility. Ecology teaches its practitioners a bit more humility. Any ecological model that contains a singular point is assumed to break down as that particular value of the independent variable is approached. It is patently assumed that some unspecified phenomenon more characteristic of the scale of events in the neighborhood of the singularity will come to dominate affairs there. Under the lens of the hierarchical view, the world appears not uniformly continuous, but rather "granular." The effects of events occurring at any one level are assumed to have diminishingly less impact at levels further removed.

Not Quite a Mechanism

Abandoning universality seems at first like a formula for disaster. What with different principles operant at different scales, the picture appears to grow intractable. But upon further reflection it should become clear that the hierarchical perspective actually offers the possibility to contain the consequences of anomalies or novel, creative events within the hierarchical sphere in which they arise. By contrast, the Newtonian viewpoint, with its universal determinism, left no room whatsoever for anything truly novel to occur. The changes it dealt with, such as those of position or momenta, appear superficial in comparison to the ontic changes one sees among living systems. That is, in the hierarchical world something truly new can happen at a particular level without causing events at distant scales to run amok.

Darwin hewed closely to the Newtonian sanctions of his time. It was therefore a looming catastrophe for evolutionary theory when Mendel purported that variation and heritability were discrete, not continuous in nature. For with discontinuity comes unpredictability and history. The much reputed "grand synthesis" by Ronald Fisher et al. sought to stem the hemorrhaging of belief in Darwinian notions by assuming that all discontinuities were confined to the netherworld of genomic events, where they occurred in complete isolation from each other. Fisher's synthesis was an exact parallel to the earlier attempt by Boltzman and Gibbs to reconcile chance with newtonian dynamics in what came to be called "statistical mechanics" (Depew and Weber 1994).

It appears to be belief and not evidence that confines chance and stochastic behavior to minuscule scales. For, if all events above the physical scale of genomes are deterministic, then one should be able to map unambiguously from any changes in genomes to corresponding manifestations at the macroscale of the phenomes. It was to test exactly this hypothesis that Sidney Brenner and numerous colleagues expended millions of dollars and years of labor (Lewin 1984). Perhaps the most remarkable thing to emerge from this grand endeavor was the courage of the project leader, who ultimately declared,

> An understanding of how the information encoded in the genes relates to the means by which cells assemble themselves into an organism...still remains elusive...At the beginning it was said that the answer to the understanding of development was going to come from a knowledge of the molecular mechanisms of gene control... [But] the molecular mechanisms look boringly simple, and they do not tell us what we want to know. *We have to try to discover the principles of organization, how lots of things are put together in the same place.* [Italics added.]

In a vague way Brenner is urging that we reconsider the nature of causality. In fact, some very influential thinkers, such as Charles S. Peirce, long ago have advocated the need to abandon causal closure. In doing so they were not merely suggesting that the ancient notions of formal and final causes be rehabilitated (as has been recommended by Rosen [1985]). None other than Karl R. Popper, whom many regard as a conservative figure in the philosophy of science, has stated unequivocally that we need to forge a totally new perspective on causality, if we are to achieve an "evolutionary theory of knowledge."

To be more specific, Popper (1959) claims we inhabit an "open" universe—that chance is not just a matter of our inability to see things in sufficient detail. Rather, indeterminacy is a basic feature of the very nature of our universe. It exists at *all* scales—not just the submolecular. For this reason, Popper says we need to generalize our notion of "force" to account for such indeterminacy. Forces deal with determinacy: if A, then B—no exceptions! What we are more likely to see under real-world conditions, away from the laboratory or the vacuum of space, Popper (1990) suggests, are the "propensities" for events to follow one another: If A, then probably B. But the way remains open for C, D, or E at times to follow A. Popper hints that his propensities are related to (but not necessarily identical to) conditional probabilities. Thus, if A and B are related to each other in Newtonian fashion, then $p(B|A) = 1$. But under more general conditions, $p(B|A) < 1$. Furthermore, $p(C|A)$, $p(D|A)$, etc. > 0.

Popper highlights two other features of propensities: (1) They may change with time. (2) Only forces exist in isolation; propensities do not. In particular, propensities exist in proximity to and interact with other propensities. The end result is what we call development or evolution. Changes of this nature are beyond the capabilities of Newtonian description.

What Popper does not provide is a concrete way to quantify, and therefore make operational, his notion of propensity. He states only, "We need to develop a calculus of conditional probabilities." So we are left to ask what can happen when lots of propensities "are put together in the same place", to use Brenner's words? How does one quantify the result? In what way do conditional probabilities enter the calculus? How does the idea of propensity relate to the Aristotelian concepts of formal and final causes?

We begin our investigation into these issues first by concentrating on what might happen when lots of processes occur in proximity. To do this we take a lead from Odum (1959) and consider all qualitative combinations of how any two processes may affect each other. Thus, process A might affect B by enhancing the latter (+), decrementing it (-), or it could have no effect whatsoever on B (0). Conversely, B could affect A in the same three ways. Hence, there are nine possibilities for how A and B can interact: (+,+), (+,-), (+,0), (-,-), (-,+), (-,0), (0,0), (0,+), and (0,-). We wish to argue that, in an open universe, the first combination, mutualism (+,+), contributes toward the organization of an ensemble of life processes in ways quite different from the other possibilities; and, furthermore, that it induces the ensemble to exhibit properties that are decidedly nonmechanical in nature. Mutualism is the glue that binds the answers to our list of questions into a unitary description of development.

When mutualisms exist among more than two processes, the resulting constellation of interactions has been characterized as "autocatalysis." A three-component example of autocatalysis is illustrated schematically (Figure 8.2). The plus sign near the box labeled B indicates that process A has a propensity to enhance process B. B, for its part, exerts a propensity for C to grow, and C, in its turn, for A to increase in magnitude. Indirectly, the action of A has a propensity to increase its own rate and extent—whence "autocatalysis."

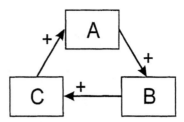

FIGURE 8.2
Schematic of a three-component autocatalytic cycle.

A convenient example of autocatalysis in ecology is the community of processes connected with the growth of macrophytes of the genus *Utricularia*, or the bladderwort family (Bosserman 1979). Species of this genus inhabit freshwater lakes over much of the world, and are abundant especially in subtropical, nutrient-poor lakes and wetlands. A schematic of the species *U. floridana*, common to karst lakes in central Florida, is depicted (Figure 8.3). Although *Utricularia* plants sometimes are anchored to lake bottoms, they do not possess feeder roots that draw nutrients from the sediments. Rather, they absorb their sustenance directly from the surrounding water. One may identify the growth of the filamentous stems and leaves of *Utricularia* into the water column with process A mentioned above.

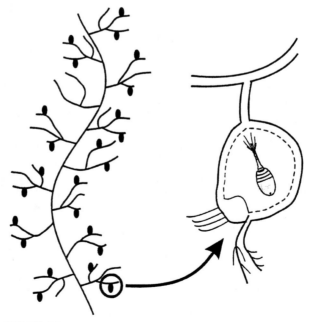

FIGURE 8.3
Rough sketch of a "leaf" of the species *Utricularia floridana*.

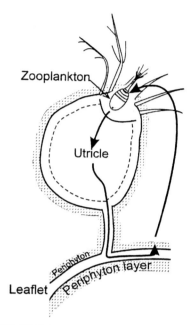

FIGURE 8.4
An autocatalytic cycle in *Utricularia* systems.

Upon the leaves of the bladderworts invariably grows a film of bacteria, diatoms, and blue-green algae that collectively are known as periphyton. Bladderworts are never found in the wild without their accoutrement of periphyton. Apparently, the only way to raise *Utricularia* without its film of algae is to grow its seeds in a sterile medium (Bosserman 1979). Suppose we identify process B with the growth of the periphyton community. It is clear, then, that bladderworts provide an areal substrate which the periphyton species (not being well adapted to growing in the pelagic, or free-floating mode) need to grow.

Now enters component C in the form of a community of small, almost microscopic (about 0.1-mm) motile animals, collectively known as "zooplankton," which feed on the periphyton film. These zooplankton can be from any number of genera of cladocerae (water fleas), copepods (other microcrustacea), rotifers, and ciliates (multicelled animals with hairlike cilia used in feeding). In the process of feeding on the periphyton film, these small animals occasionally bump into hairs attached to one end of the small bladders, or utrica, that give the bladderwort its family name. When moved, these trigger hairs open a hole in the end of the bladder, the inside of which is maintained by the plant at negative osmotic pressure with respect to the surrounding water. The result is that the animal is sucked into the bladder, and the opening quickly closes behind it. Although the animal is not digested inside the bladder, it does decompose, slowly releasing nutrients that can be absorbed by the surrounding bladder walls. The cycle (Figure 8.2) is now complete (Figure 8.4).

Because the example of indirect mutualism provided by *Utricularia* is so colorful, it becomes all too easy to become distracted by the mechanical-like details of how it, or any other example of mutualism, operates. The temptation naturally arises to identify such autocatalysis as a "mechanism," as it is referred to in the field of chemistry. In the closed world of mechanical-like reactions and fixed chemical forms, such characterization of autocatalysis is legitimate. It becomes highly inappropriate, however, in an open universe, such as a karst lake, where connections are probabilistic and forms can exhibit variation. There autocatalysis can exhibit behaviors that are decidedly nonmechanical. In fact, autocatalysis under open conditions can exhibit any or all of eight characteristics, which, taken together, separate the process from conventional mechanical phenomena (Ulanowicz 1997).

To begin with, autocatalytic loops are (1) *growth enhancing*. An increment in the activity of any member engenders greater activities in all other elements. The feedback configuration results in an increase in the aggregate activity of all members engaged in autocatalysis over what it would be if the compartments were decoupled. In addition, there is the (2) *selection pressure* which the overall autocatalytic form exerts upon its components. For example, if a random change should occur in the behavior of one member that either makes it more sensitive to catalysis by the preceding element or accelerates its catalytic influence upon the next compartment, then the effects of such alteration will return to the starting compartment as a reinforcement of the new behavior. The opposite is also true. Should a change in the behavior of an element either make it less sensitive to catalysis by its instigator or diminish the effect it has upon the next in line, then even less stimulus will be returned via the loop.

Unlike Newtonian forces, which always act in equal and opposite directions, the selection pressure associated with autocatalysis has the effect of (3) *breaking symmetry*. Autocatalytic configurations impart a definite sense (direction) to the behaviors of systems in which they appear. They tend to ratchet all participants toward ever greater levels of performance.

Perhaps the most intriguing of all attributes of autocatalytic systems is the way they affect transfers of material and energy between their components and the rest of the world. Figure 8.2 does not portray such exchanges, which generally include the import of substances with higher exergy (available energy) and the export of degraded compounds and heat. What is not immediately obvious is that the autocatalytic configuration actively recruits more material and energy into itself. Suppose, for example, that some arbitrary change happens to increase the rate at which materials and exergy are brought into a particular compartment. This event would enhance the ability of that compartment to catalyze the downstream component, and the change eventually would be rewarded. Conversely, any change decreasing the intake of exergy by a participant would ratchet down activity throughout the loop.

The same argument applies to every member of the loop, so that the overall effect is one of (4) *centripetality,* to use a term coined by Sir Isaac Newton.

By its very nature autocatalysis is prone to (5) *induce competition,* not merely among different properties of components (as discussed above under selection pressure), but its very material and (where applicable) mechanical constituents are themselves prone to replacement by the active agency of the larger system. For example, suppose A, B, and C are three sequential elements comprising an autocatalytic loop (Figure 8.2), and that some new element D: (a) appears by happenstance, (b) is more sensitive to catalysis by A, and (c) provides greater enhancement to the activity of C than does B. Then D either will grow to overshadow the role of B in the loop, or will displace it altogether. In like manner one can argue that C could be replaced by some other component E, and A by F, so that the final configuration D-E-F would contain none of the original elements. It is important to notice in this case that the characteristic time (duration) of the larger autocatalytic form is longer than that of its constituents.

The appearance of centripetality and the persistence of form beyond constituents make it difficult to maintain hope for a strictly reductionist, analytical approach to describing organic systems. Although the system requires material and mechanical elements, it is evident that some behaviors, especially those on a longer time scale, are, to a degree, (6) *autonomous* of lower level events (Allen and Starr 1982). Attempts to predict the course of an autocatalytic configuration by ontological reduction to material constituents and mechanical operation are, accordingly, doomed over the long run to failure.

It is important to note that the autonomy of a system may not be apparent at all scales. If one's field of view does not include all the members of an autocatalytic loop, the system will appear linear in nature. One can, in this case, seem to identify an initial cause and a final result. The subsystem can appear wholly mechanical in its behavior. For example the phycologist who concentrates on identifying the genera of periphyton found on *Utricularia* leaves would be unlikely to discover the unusual feedback dynamics inherent in this community. Once the observer expands the scale of observation enough to encompass all members of the loop, however, then autocatalytic behavior with its attendant centripetality, persistence, and autonomy (7) *emerges* as a consequence of this wider vision.

Finally, it should be noted that an autocatalytic loop is itself a kinetic form, so that any agency it may exert will appear as a (8) *formal* cause in the sense of Aristotle.

One may summarize these various effects of autocatalysis in thermodynamic terms as either extensive or intensive in nature. Extensive system properties pertain to the size of a system, whereas intensive attributes refer to those qualities that are structural and independent of system size. Thus, growth enhancement is decidedly extensive. The remaining properties are intensive and serve to prune from the kinetic structure of the system those pathways that less effectively participate in autocatalysis. The augmented flow activity is progressively constrained to flow along the (autocatalytically)

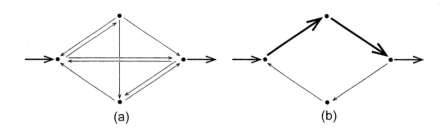

FIGURE 8.5
Schematic depiction of the effects that autocatalysis exerts upon networks. (a) Before;
(b) After.

more efficient routes as the system "develops." The combination of extensive increase in system activity and intensive system development is depicted schematically (Figure 8.5).

Quantifying Kinetic Constraints

Properties of systems do not truly enter scientific dialog until they have been made fully operational. That is, until it becomes possible to quantify and measure the effects of autocatalysis upon a system, all talk about organization and development in living systems remains purely speculative. In order to ensure that at least some identifiable cause (material causality) will always remain explicit in our system description, we choose to quantify only those relationships between compartments that can be measured in terms of a palpable exchange of some material constituent, such as carbon, energy, nitrogen, or phosphorus. No one is assuming that these exchanges are the only ones, nor even the most important ones, that transpire in the system and give it its form. Whatever the actual natures of the causal events, however, their effects will be manifested as changes in the material transactions among the members of the community.

Accordingly, we define T_{ij} as the amount of the chosen medium that is donated by prey i to predator j per unit space per unit time. As explained above, not all exchanges are among the n system components. Exogenous transfers also must be accounted. Thus, we will assume that imports from outside the system originate in taxon 0 (zero). Furthermore, we will distinguish two types of outputs from the system: material that is exported in a form still usable to some other system of comparable size will be assumed to flow to component n + 1, whereas material that has been reduced to some marginally useful "ground state" (e.g., carbon dioxide) will be accounted as flowing to compartment n + 2.

The material assumption and the exhaustive accounting scheme just described make possible the quantification of both the extensive and intensive effects of autocatalysis. To quantify the extensive changes is almost trivial. By a change in system activity is meant any fluctuation in the aggregate of all transactions currently underway. In economic theory this sum is called the "total system throughput" and will appear as

$$T_{..} = \sum_{i,j} T_{ij} \tag{1}$$

where a dot in place of a subscript indicates that particular subscript has been summed over all components from 0 to $n + 2$. It follows that any increase in the level of system activity will be reflected as a rise in $T_{..}$.

Changes in the intensive character of a system are somewhat more difficult to quantify, but the effort is crucial, because in doing so we are addressing the crux of this essay—the quantification of system constraints. We begin this task by first turning our attention to the lack of constraint, or the *indeterminacy* of event i. Such indeterminacy was quantified more than a century ago by Ludwig von Boltzmann

$$S_i = -k \log p(A_i), \tag{2}$$

where $p(A_i)$ is the probability of event A_i happening, k is a scalar constant, and S_i is the *(a priori)* indeterminacy associated with i. Sometimes S_i is called the surprisal of A_i, because, if the probability of A_i is very small (near zero), we become very surprised when it does occur (S_i is large.)

We now try to follow Brenner's advice and quantify what happens when lots of things are put together. Specifically, we ask "How is the indeterminacy of A_i changed whenever event B_j has just occurred?" Or, in terms that pertain more to this essay, "By how much does the presence of B_j *constrain* event A_i?" By "constrain" we mean "decrease the indeterminacy" of A_i. When B_j precedes A_i, any constraint that it exerts upon the latter will be reflected by a change in probability that A_i will occur. This altered probability is nothing other than the conditional probability of A_i, given B_j. Thus, indeterminacy has been diminished to

$$S_{ij} = -k \log p\left(A_i \middle| B_j\right), \tag{3}$$

where S_{ij} is now the *a posteriori* indeterminacy of A_i given B_j. Accordingly, the reduction in indeterminacy that is calculated by subtracting S_{ij} from S_i becomes a measure of the constraint that B_j exerts on A_i. Remembering that the negative of a logarithm is equal to the logarithm of the reciprocal of its argument, and that the difference between two logarithms is the same as the logarithm of the quotient of the two arguments, we find that $S_i - S_{ij}$ becomes

$$S_i = S_{ij} = k \log\left[p\left(A_i|B_j\right)\middle/p\left(A_i\right)\right].$$ (4)

Here we note that Bayes' Theorem allows one to calculate $p(A_i \mid B_j)$ as the quotient of $p(A_i,B_j)$ by $p(B_j)$, where $p(A_i,B_j)$ is the joint probability that Ai and B_j occur in combination. Whence, (4) may be rewritten in the more symmetrical form,

$$S_i - S_{ij} = k \log\left[p\left(A_i, B_j\right)\middle/p\left(A_i\right)p\left(B_j\right)\right]$$ (5)

Because A_i and B_j are any arbitrary pair of events, it becomes an easy matter to calculate the average amount of constraint that all system elements exert upon each other. One simply multiplies Equation 5 for each combination i and j by the probability that A_i and B_j co-occur and sums over all combinations of i and j. The resulting "average mutual constraint" looks like

$$k \sum_{i,j} p\left(A_i, B_j\right) \log\left[p\left(A_i, B_j\right)\middle/p\left(A_i\right)p\left(B_j\right)\right]$$ (6)

To make Equation 6 operational it remains only to estimate the three probabilities in terms of measured quantities. If one regards the trophic exchanges as entries in an events matrix, then it would follow immediately that:

$$p\left(A_i, B_j\right) \sim T_{ij}/T_{..}$$

$$p\left(A_i\right) \sim T_i/T_{..}$$ (7)

$$p\left(B_j\right) \sim T_j/T_{..}$$

Substituting Equation 7 into Equation 6 yields

$$AMI = k \sum_{i,j} \left(T_{ij}/T_{..}\right) \log\left[T_{ij}T_{..}/T_i T_{.j}\right],$$ (8)

where AMI is the "average mutual information" of information theory. ("Information" and "constraint" are interchangeable in information theory.) Two familiar results from information theory are that AMI is intrinsically non-negative and that it is bounded from above by the index

$$H = -k \sum_{i,j} \left(T_{ij}/T_{..}\right) \log\left(T_{ij}/T_{..}\right)$$ (9)

where H is the overall indeterminacy of the flow structure (Ulanowicz and Norden 1990).

The reader is encouraged to apply Equation 8 to any variety of flow network configurations to convince oneself that the AMI accurately measures the intensive change in kinetic structure from that in Figure 8.5a to the one in Figure 8.5b. A hypothetical example is given (Figure 8.6).

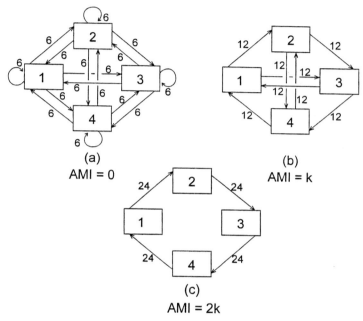

FIGURE 8.6
Three hypothetical networks illustrating how average mutual information (AMI) increases with the degree of network pruning.

The results of the calculations (Figure 8.6) are presented in terms of units of k, which have yet to be specified. The usual convention in information theory is to choose a base for the logarithms (either 2, e, or 10), set k = 1, and call the resulting units "bits," "napiers," or "hartleys," respectively. Doing thusly would leave us with two separate measures for the extensive and intensive attributes of flow networks. Both properties are strongly influenced, however, by a single process—the autocatalysis. We therefore emphasize the unitary origin of changes in both aspects by following the advice of Tribus and McIrvine (1971); we use the scalar factor k to impart physical dimensions to our measure of constraint. Setting k = T.. in Equation 8 gives

$$A = \sum_{i,j} T_{ij} \log\left(T_{ij} T_{..} / T_{i.} T_{.j}\right) \qquad (10)$$

where the scaled index, A, is renamed the system "ascendency." It is an amalgamated measure of the tendency for a system to increase in both activity and structure (constraint) via internal autocatalysis.

We note that the ascendency is fully operational, as the formula for A consists entirely of measurable quantities. That is, for each and every fully quantified network of trophic exchanges, one may calculate a unique value of A. After one evaluates a number of networks in this fashion, it becomes apparent that certain network attributes are associated with increases in A. These include: (1) specialization, (2) speciation, (3) internalization, and (4) cycling. These same properties, however, are recognized as the broad categories that group the 24 attributes identified by Odum (1969) to characterize the late successional stages of a developing ecosystem. One is prompted, then, to suggest as a phenomenological principle:

> In the absence of major perturbations, ecosystems naturally tend towards configurations of ever-greater ascendency.

Before applying ascendency to spatially heterogeneous ecosystems, it is important to stress two points. The first is that increasing ascendency is only one half of the development story. Ascendency encompasses all that is efficient and productive about the network configuration. Although we have cited the inclination for a system to progress in this direction, it cannot be overemphasized that this tendency is often desultory and at times could culminate in the destruction of the system. For increasing ascendency tells only what happens in the absence of relatively heavy perturbations. Should the system progress too far in the direction of increasing efficiency, it will become "brittle" (Holling 1986) and lack the flexibility to adapt whenever the system is impacted by novel disturbances.

Fortunately, one can readily construct a complement to ascendency using quantities already defined. One recalls that the average mutual information was bounded by the Quantity 9 (which, effectively, quantifies the diversity of system flows). This indeterminacy may be scaled by $T_.$ in exactly the same manner as was done to the AMI. The result, called the system capacity, becomes an upper bound on the ascendency. The amount by which this capacity exceeds the ascendency is called the system "overhead", and it quantifies all the inefficient, indeterminate, and diffuse processes that remain in the system. The capacity also includes the degrees of freedom inherent that the system can use to reconfigure in the aftermath of a significant perturbation. Without sufficient overhead, a system is doomed to death or major collapse.

The second issue concerns the role of biomasses or stocks in system development. The ascendency as formulated above contains no explicit mention of taxon bemuses. Yet classical dynamics suggest that stocks cannot be entirely ignored. Fortunately, a way was recently discovered to incorporate stocks of components into the ascendency in a manner that accords with the requirements of information and probability theories (Ulanowicz and Abarca–Arenas 1997). The new formulation for the ascendency is

$$A = \sum_{i,j} T_{ij} \log\left[T_{ij} B^2 / B_i B_j T \right] \tag{11}$$

where B_i is the biomass of component i. Definition 11 will be employed to calculate the ascendency in the remainder of this paper.

Landscapes of Flows

If ascendency theory as presented here should seem a bit abstract, the reader should find compensation in knowing that abstractness carries with it broad generality. For example, the flow T_{ij} was defined as the trophic exchange from prey i to predator j. It could just as well represent the movement of a given amount of a species from spatial position i to location j. Similarly, B_i could represent the density of the given population at location i. When one substitutes these new variables into Equation 11, the ascendency that results now applies to the migration of the given population over the landscape. The ascendency hypothesis as it pertains to migration translates into:

> In the absence of massive perturbations, the populations of an ecosystem distribute themselves across a landscape in a way that leads progressively to higher system ascendencies.

(It should be noted in passing that it is likewise possible to apply the ascendency measure to several populations migrating across a landscape while simultaneously engaging in trophic interactions at each point in space ([Ulanowicz 1997].)

It is the utility of applying ascendency-like variables to biotic movements across landscapes that we wish to explore in the remainder of this essay. In the interest of simplicity, it will help if we keep the landscape rather simplistic. Toward this end, we will consider a 10×10 grid of spatial elements upon which we will run five separate models in the manner of cellular automata (CA). The elements of the two-dimensional spatial array will be numbered sequentially by a single running index (Figure 8.7). To simplify the boundary conditions at the edges of the landscape, we shall assume that the edges "wrap around" in both the horizontal and vertical directions. That is, transport beyond the "eastern" (right-hand) edge of the domain will feed into the western margin, as shown in the figure.

The first model simulates nearest neighbor diffusion. Material or organisms in adjacent cells exchange material across their common boundary at a

	91	92	93	94	95	96	97	98	99	100	
10	1	2	3	4	5	6	7	8	9	10	1
20	11	12	13	14	15	16	17	18	19	20	11
30	21	22	23	24	25	26	27	28	29	30	21
40	31	32	33	34	35	36	37	38	39	40	31
50	41	42	43	44	45	46	47	48	49	50	41
60	51	52	53	54	55	56	57	58	58	60	51
70	61	62	63	64	65	66	67	68	69	70	61
80	71	72	73	74	75	76	77	78	79	80	71
90	81	82	83	84	85	86	87	88	89	90	81
100	91	92	93	94	95	96	97	98	99	100	91
	1	2	3	4	5	6	7	8	9	10	

FIGURE 8.7
The numbering scheme used in a 10 × 10 gridwork of landscape elements. Marginal rows and columns illustrate the "wrap-around" boundary conditions.

rate that is proportional to the difference in population density or biomass across that same boundary. Thus, for any time step we calculate in the horizontal (west–east) direction,

$$T_{i-1,i} = D\left(B_{i-1} - B_i\right)$$
$$T_{i,i+1} = D\left(B_i - B_{i+1}\right)$$

(12a)

and in the vertical (north–south) direction,

$$T_{i-10,i} = D\left(B_{i-10} - B_i\right)$$
$$T_{i,i+10} = D\left(B_i - B_{i+10}\right)$$

(12b)

(where D is a constant coefficient of exchange). The biomasses at all locations are thereafter incremented in the fashion

$$B_i* = B_i + T_{i-1,i} - T_{i,i+1} + T_{i-10,i} - T_{i,i+10},$$

(13)

where B_i* becomes the biomass at gridpoint i during the next iteration.

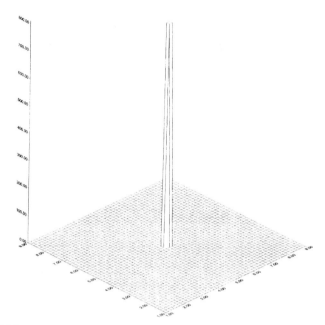

FIGURE 8.8A
Animal density profiles (arbitrary units) for a random-walk dispersion. At the beginning.

This simulation of diffusion also approximates a random-walk migration scenario. We begin the simulation with a given quantity of organisms concentrated in a single cell at the center (Figure 8.8A). For the chosen value of the diffusion parameter ($D = 0.1$), dispersion across the landscape is quite rapid (Figure 8.8B and Figure 8.8C), and a virtually uniform dispersion is reached by timestep 100. As one might expect, the system ascendency for this scenario dies off in approximately exponential fashion (Figures 8.9).

To examine the dynamics in somewhat greater detail, we wish to plot how the full ascendency is distributed across the landscape. The reader will recallthat Formula 11 involves a double summation. To gauge the contribution to the ascendency made by all organisms arriving at a given gridpoint, one simply sums over the first index while leaving the other one free. That is, for each gridpoint j, one may calculate

$$A_j = \sum_i T_{ij} \log\left(T_{ij} B_{.}^2 / B_i B_j T_{..}\right), \tag{14}$$

where A_j is the contribution made by all organisms at point j towards the full landscape ascendency. Figure 8.10 shows the distribution of the landscape ascendency for the diffusion model at timestep 6. The distribution resembles an eroded volcanic crater. (The humps along the rim are artifacts of the small

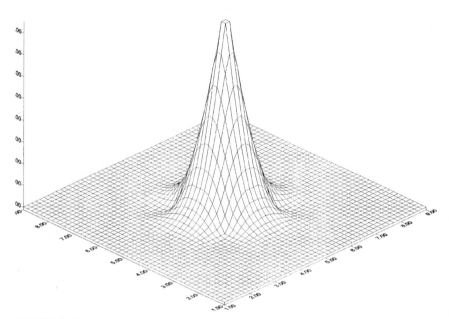

FIGURE 8.8B
Animal density profiles (arbitrary units) for a random- walk dispersion. After the first time step.

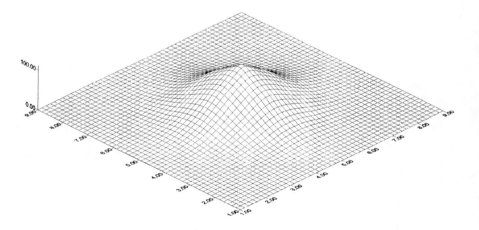

FIGURE 8.8C
Animal density profiles (arbitrary units) for a random- walk dispersion. After 6 time steps.

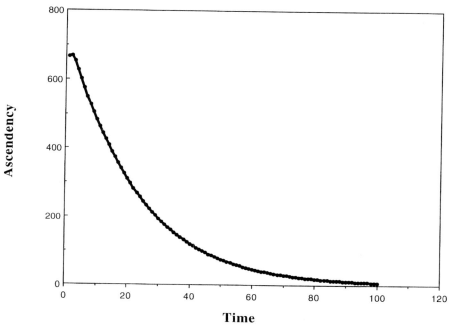

FIGURE 8.9
Change in total landscape ascendency during the random-walk dispersion scenario.

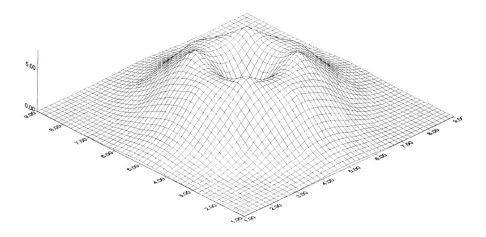

FIGURE 8.10
Distribution of the spatial components of ascendency after timestep 6 of the random-walk scenario.

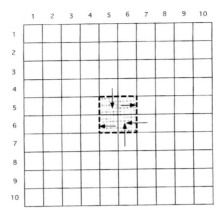

FIGURE 8.11
"Maxwell's Box" scenario for animal aggregation.

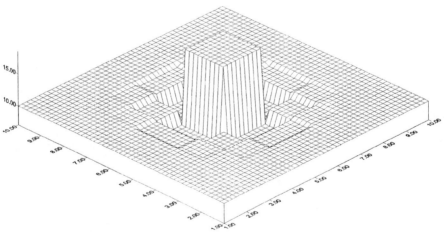

FIGURE 8.12A
Animal density profiles (arbitrary units) for Maxwell's Box aggregation. After 5 timesteps.

number of gridpoints in the landscape.) The key thing to notice is that the important action is not occurring at the center (where there is greatest density, but little diffusion), but at a certain distance from the center, where biomass gradients are steepest and migration strongest.

As old as the myth of Pandora's Box is the notion that some processes are irreversible. It is not surprising, therefore, to find that one cannot readily run the dispersion model in reverse. An approximation to such a reversal we shall call "Maxwell's Box." Maxwell's Box is an area of four grid cells at the center of the landscape (Figure 8.11). It is called Maxwell's Box in analogy to the famous Maxwellian Demon, which was a hypothetical being stationed at a pinhole in a partition that separates two chambers that initially are filled

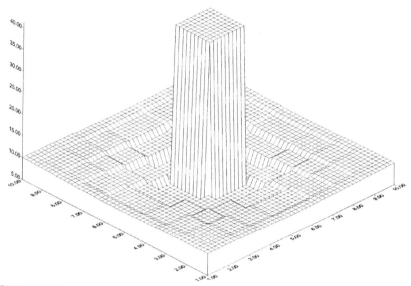

FIGURE 8.12B
Animal density profiles (arbitrary units) for Maxwell's Box aggregatioN. After 25 timesteps.

with a mixture of two gases, say A and B. The demon operated a frictionless, massless trapdoor over the hole, which he would open if a molecule of B approached from the left or if a molecule of A came from the right. Otherwise, he would leave the flap closed. Eventually, the gases would separate—A into the left chamber and B into the right in ostensible contradiction to the Second Law of Thermodynamics. In our analog, if an animal wanders into Maxwell's Box, it does not leave. The situation is analogous to animals doing a random-walk search for suitable habitat (the box). Once they find it, they stay put. Eventually, most of the animals wind up in the box (Figure 8.12A and Figure 8.12B).

At first thought, one might anticipate a logistic-like increase in system ascendency over time, i.e., the reverse of Figure 8.9. Instead, the ascendency rises for about 30 timesteps, then goes into a slow decline (Figure 8.13.) The initial rise is due primarily to an increase in mass segregation that is occurring over the landscape. The slow decline results from the gradual decline in activity as most of the animals end up in the box. The distribution of ascendency over the landscape at timestep 25 is rather unremarkable—a hill in the middle of the landscape, similar to the form in Figure 8.8C.

For the third scenario we impose a uniform migration of animals from north to south. This is accomplished by amending Equation 12b to read:

$$T_{i-10,i} = UB_{i-10} + D\left(B_{i-10} - B_i\right)$$

$$T_{i,i+10} = UB_i + D\left(B_i - B_{i+10}\right)$$

(12c)

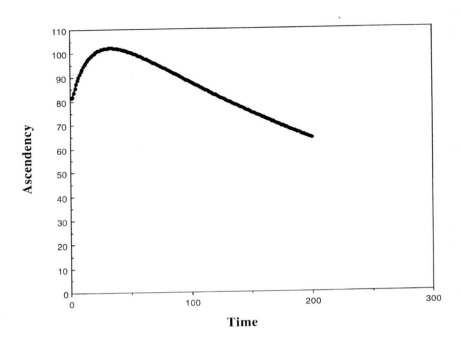

FIGURE 8.13
Change in total landscape ascendency during the course of Maxwell's Box aggregation.

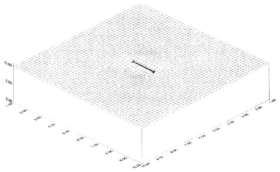

FIGURE 8.14A
Development of animal distributions along a migratory stream that encounters a crosstream barrier. After 2 timesteps.

where U is a constant rate of migration (or advection, as the case may be). In contrast to the endpoint of our first diffusion scenario (a uniform density across the landscape), the uniform flow possesses both a preferred direction and an observable amount of net migration activity. These attributes give rise to a nonzero ascendency (256.8 flow bits) and an appreciable total system

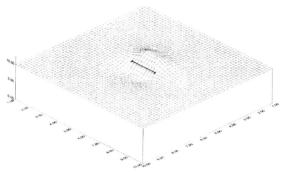

FIGURE 8.14B
Development of animal distributions along a migratory stream that encounters a crosstream barrier. After 10 timesteps.

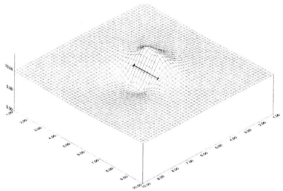

FIGURE 8.14C
Development of animal distributions along a migratory stream that encounters a crosstream barrier. After 100 timesteps.

throughput (40.5 flow units), respectively. One may say, therefore, that the flow field possesses 256.8 flow bits of organization.

With the fourth scenario we address directly the title of this chapter. In the very middle of the uniform flow field we place an impermeable barrier two gridpoints wide. As might be expected, organisms begin to accumulate upstream and become depleted downstream of the barrier (Figure 8.14A and Figure 8.14B). Diffusion in the east-west directions eventually brings the system to a steady-state after about 100 timesteps (Figure 8.14C). Isopleths of animal density reveal the regions of accumulation and depletion, as well as a faint "bow-wake" forward and aft of the barrier itself (Figure 8.15). The migratory flow field reveals a parting of the migration stream around the barrier (Figure 8.16). The accompanying steady-state distribution of the landscape ascendency (Figure 8.17). It resembles a valley that is perpendicular to the barrier, flanked on both sides by two ridges that parallel the

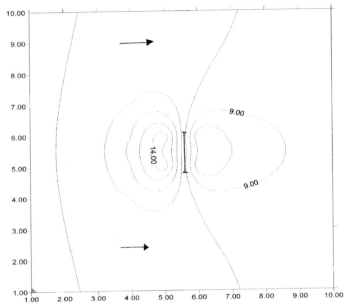

FIGURE 8.15
Isopleths of animal densities (after 100 timesteps) in relation to the barrier imposed upon a uniform migratory stream.

migratory stream. The ridges are highest just downstream of the barrier, whereas the greatest flow magnitudes appear just upstream and to the sides of the barrier.

Comparison of the system properties of the undisturbed stream flow with those corresponding to flow around the barrier is likewise revealing. Interestingly, more total flow occurs with the barrier in place (49.88 flow units) than in the unimpeded situation (40.50). This increase is an artifact of both the particular boundary conditions and the lack of any explicit resistance term in the CA scheme. As a consequence of the augmented flow, the ascendency increases from 256.8 flow bits in the uniform migration to 322.3 with the barrier. The mutual information of the flow field (A/T) increases from 6.341 bits without the barrier to 6.462 bits with the constraint. One may conclude, therefore, that the barrier constraint adds 65.5 flow bits of ascendency to the dynamics of the system and 0.121 bits to its organization.

As a final exercise we compare the organization inherent in purely random movements across the whole landscape with that pertaining to the same amount of migration between two specific points in the field. In the former simulation, the origins and destinations for 10,000 "flights" were chosen at random from the entire field of 100 gridpoints. (This is different than the random walk considered earlier, where transitions were confined to nearest neighbors.) The first 40 of these "flights" are graphed in Figure 8.18. In the highly constrained migration, 5000 flights occurred from gridpoint 19 to cell

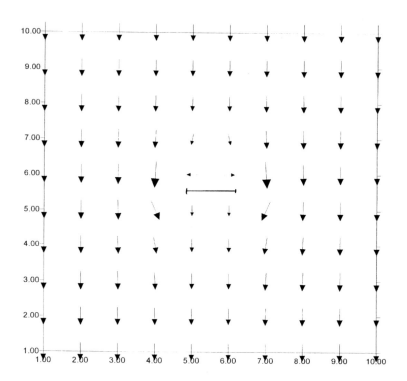

FIGURE 8.16
Vector field of animal movements around a barrier (corresponding to Figure 8.14C).

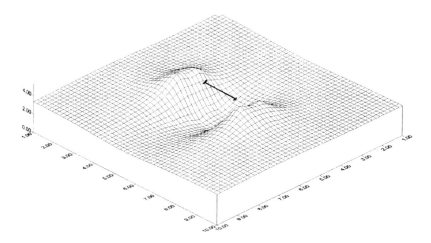

FIGURE 8.17
Distribution of the ascendency components across the landscape after timestep 100 of the "barrier" scenario.

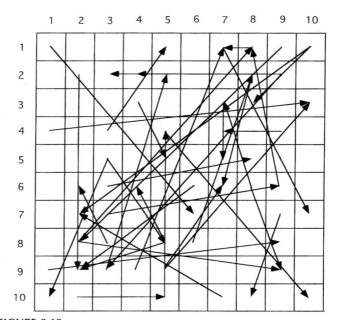

FIGURE 8.18
The initial 40 (of 10,000) random "flights" across a landscape.

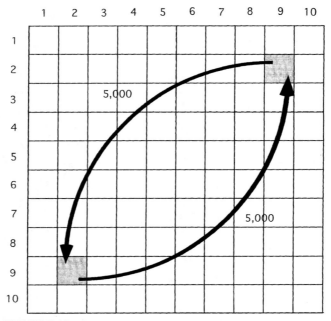

FIGURE 8.19
Schematic of 10,000 highly determinate migrations between two specific gridpoints in a landscape.

82 and the same number in the reverse direction (Figure 8.19). The latter scenario resembles the migratory flights of certain bird populations between two areas of suitable habitat situated among a sea of unsuitable locales.

In both cases the total activity was the same—10,000 flow units. In the random flight scenario these transitions resulted in 9,517 flow bits of ascendency distributed more or less evenly over the landscape. In the highly constrained situation the same amount of activity yielded 122,900 flow bits of ascendency, concentrated at the two sites of suitable habitat. The AMI in the first case was 0.9517 bits, whereas that for the latter was 12.29 bits. Although we have no knowledge concerning the details of the constraints operating in the second case, we nonetheless can conclude that they contribute almost 113,400 flow bits to the dynamics of the system. Furthermore, almost 13 times as much information is associated with the very organized process of migration depicted in Figure 8.17.

Conclusions

Coming to terms with an indeterminate world requires that we explore new methods for quantifying natural phenomena. Earlier perspectives, which view phenomena either as determinate and continuous or wholly stochastic, must be supplemented by a "calculus of conditional probabilities." Such a calculus has already been provided by information theory. It is counterproductive to consider that information theory has only narrow application to problems related to communications theory. Rather, it is universally applicable whenever indeterminate phenomena become significant—which encompasses most of the life and social sciences.

In particular, the information indices that have been used to quantify trophic constraints among an ecological community can likewise quantify the organizational constraints operating on populations of animals that move across a landscape. Thus, the hypothesis of increasing ascendency might pertain to landscape ecology as well. Because the hypothesis is cast in terms that can be quantified using data on population distributions and migrations, it can be made operational and thus subject to falsification.

From a more practical point of view, the distribution of ecosystem ascendency over the landscape can be calculated to identify the "hot spots" where the most significant quantitative events are occurring.

Finally, it should be noted that the theory of ascendency treats a spatially distributed ecosystem as a grand process, rather than as a frozen picture of the world. It would seem that such transition is necessary if landscape ecology is to become a viable life science.

Acknowledgments

The author would like to thank Dr. Lawrence Harris for suggesting this endeavor and Dr. James Sanderson for the welcome encouragement he gave this writer to make time for the formidable tasks of creating and running the models. Mr. James Hagy and Dr. Frances Rohland helped tutor the author in graphical software. Mrs. Jeri Pharis was most helpful in putting the manuscript into decent final shape.

The author was supported by the USGS Biological Resources Division as part of their Across Trophic Levels System Simulation (ATLSS) project (1445-CA09-0093) and by the Multiscale Experimental Ecosystem Research Center (MEERC), funded by the U.S. EPA (Contract R819640).

9

Land Use in America: The Forgotten Agenda

John F. Turner and Jason Rylander

CONTENTS

Introduction

Take a look across America. From Boston to Baton Rouge, massive changes have taken place on the landscape and in our society. A seasoned traveler, dropped onto a commercial street anywhere in America, could scarcely tell the location from the immediate vista. A jungle of big-box retailers, discount stores, fast-food joints, and gaudy signs separated by congested roadways offer no clues to location. Every place is beginning to look like no place in particular. The homogenization of America is nearly complete.

Land use patterns viewed from the air reveal cul-de-sac subdivisions accessible only by car separated from schools, churches, and shopping spread out from decaying cities like strands of a giant spider web. Office parks and factories isolated by tremendous parking lots dot the countryside. Giant malls and business centers straddle the exit ramps of wide interstates where cars are lined bumper to bumper. Residential areas are secured from the rest of us

1-56670-368-9/00/$0.00+$.50

and defy any sense of community. Cities and towns blend for tens of miles into what is left of the country. Green spaces are fragmented. Only a remnant of natural spaces remain intact. In Florida, for instance, residential tracts are secured behind walls that defy any sense of community.

In America powerful economic and demographic forces are at work. Population growth, migration, and fractured, low-density settlement and development patterns have altered the landscape. In little more than a generation this nation has been transformed; 80% of everything built in this country was constructed in the last half century (Kunstler 1993). While much of this growth has been positive, the economic, environmental, and social costs of our current land consumption habits are now becoming increasingly apparent.

For much of America's history, expansion was a national goal. Immigrants were encouraged to settle the farthest reaches of the countryside. Land was cheap and plentiful. In a nation so vast, the notion of resource scarcity took generations to gain credibility. As early as the 1860s, however, George Perkins Marsh (1907) in his now classic work *Man and Nature* warned:

> Man has too long forgotten that the earth was given to him for usufruct alone, not for consumption, still less for profligate waste. Nature has provided against the absolute destruction of any of her elementary matter, the raw material of her works; the thunderbolt and the tornado, the most convulsive throes of even the volcano and the earthquake, being only phenomena of decomposition and recomposition. But she has left it within the power of man irreparably to derange the combinations of inorganic matter and of organic life, which through the night of aeons she had been proportioning and balancing, to prepare the earth for his habitation, when in the fullness of time his Creator should call him forth to enter into its possession.

Few listened and still fewer understood. By the turn of the 20th century many wildlife resources had been squandered. Now the U.S. is a nation of 265 million people, with a population expected to increase by half again by the year 2050. Few places are unaffected by human development.

Increasingly, our nation finds itself struggling to meet the public's competing demands for open space, wildlife, recreation, environmental quality, economic development, jobs, transportation, and housing. While it may never be possible in a democracy to meet each of these demands equitably, the tortured and fragmented way in which land use decisions are currently made all but ensures that conflict and crisis will continue to characterize environmental policy in the 21st century. It need not be so. A new land ethic must be developed, one that considers the needs of current and future generations, understands the carrying capacity of natural systems, and builds communities in which people can continue to prosper socially and economically.

Land use, we suggest, is the forgotten agenda of the environmental movement. In the past 25 years, the many environmental laws of the nation responded to one problem at a time: air or water pollution, endangered species, and waste disposal primarily through prohibitive regulatory policies that restrict private behavior. These laws have worked as stop-gap measures at best and future laws appear to offer diminishing returns.

Environmental progress in the next generation will increasingly depend on stemming the environmental costs of current land use patterns. Perhaps because "land use" is such a vague term, policymakers have difficulty grasping the linkages between the use of land and the economic, environmental, and social health of their communities. Environmental issues are traditionally debated in state and federal legislatures. Local governments and planning commissions consider land use. The next generation of environmental policy-making will require a more holistic approach that considers the impact of development on natural systems and integrates decision making across political boundaries. Policies must build on the fundamental recognition that land use decisions and environmental progress are two sides of the same coin. So long as the cumulative effects of land use decisions are ignored, we submit that environmental policy will be only marginally successful in achieving it goals.

Past Patterns

For most of the last two centuries, Americans flocked to cities seeking a better life. Since 1950, however, people have begun to flee the urban core, moving out to fast-growing areas on the periphery. This outward migration has created a doughnut-like pattern of growth on the edges and emptiness in the center. While the urbanization of America continues, in the sense that more and more people are living within metropolitan areas and suburbs, the populations of many center cities have collapsed. Of the 25 largest U.S. cities in 1950, 18 have lost population. Over the past 40 years, central Baltimore and Philadelphia have each lost more than 20% of their residents, while central Detroit declined roughly by half. St. Louis, the "Gateway to the American West," once boasted more than 850,000 people, but now has only about 400,000 residents. During the same time, suburbs across the country doubled in size, gaining 75 million people. By 1990, more Americans lived in suburbs than in cities and rural areas combined (Diamond and Noonan 1996; Jackson 1996).

The suburbanization of America has consumed a tremendous amount of land. The population of metropolitan Cleveland declined by 8% between 1970 and 1990, yet its urban land area increased by a third. Even in cities that have not declined, their geographical reach has far outpaced population growth. The population of Los Angeles grew by 45% from 1970 to 1990, but the metropolitan area of the city expanded by 300% and now equals the size of Connecticut. Metropolitan Chicago grew in population by 4% yet its developed land area expanded by 46% (Jackson 1996).

Our land use patterns affect the environment in many ways. Most notably, development pressures have significant impacts on habitat. Even where forests and wetlands are preserved, new housing and commercial developments pave over open spaces, alter water courses and runoff flows, and rearrange scenic vistas. Our land use choices also impact air quality. For example, vehicle miles traveled by the sprawling population of California have increased more than 200% in the past 2 decades as a consequence of distant suburbanization, exacerbating an already well-known smog problem in the region (Diamond and Noonan 1996). Mass transit, which is only viable at relatively high population densities, becomes increasingly impractical as people spread out across the land.

Each year, another Paris, roughly 2.2 million people, is added to the American population. New Jersey has a higher population density than Japan. If current trends continue, 80% of these people will work and settle in edge cities and areas on the metropolitan fringe. Each new single-family detached home requires public services, schools, shopping areas, extended water and sewer services, and roadways that further encroach into farmland, ranches, and open space. Coastal areas, the South, and the intermountain West face particularly acute growth challenges as more and more people, particularly retirees, migrate to these regions.

Information technology makes remote locations more accessible, and a growing number of people who now can work from their homes are also moving for the natural beauty and personal security these places afford. This phenomenon is certainly reaping disturbing consequences to the rural landscapes of the intermountain areas of the West. Without comprehensive planning to address these demographic trends, patterns of explosive growth and voracious land consumption will continue with little or no consideration of the cumulative impacts on the environment and our future well-being. To ensure a reasonable standard of living for its people and a healthy environment, the U.S. must develop more rational and productive ways to manage resources, land as well as air, water, biological systems, and people.

Unfortunately, government policies have historically exacerbated trends toward separation and expansion. Land use planning in the U.S. has traditionally been the task of local officials who have used property zoning regulations and building codes as their principal tools. Zoning, a 20th century invention, was originally intended to protect property owners from their neighbors, to ward off economic, social, or environmental damage inflicted by adjacent land use. While zoning has sometimes served these needs well, local planners have increasingly used zoning regulations to separate arbitrarily residential and commercial uses of land. As a result, the integration of shops and housing, narrow streets, and dense development that attracts admiring visitors to historic urban areas, such as the Georgetown section of Washington, D.C., is prohibited by most local codes. Yet such multi-use urban development patterns offer residents more choices in type of housing, better access and convenience, less segregation by income and class, and a greater sense of community at far less infrastructure cost.

As a whole, the U.S. land regulatory system is a failure. Multiple programs and policies are designed to address usually worthwhile goals, but are implemented in too small an area and typically without regard to the health of the region. The existing policy is one of directed chaos and is consequently oblivious to unintended consequences. As Aldo Leopold noted, "To build a better motor we tap the uttermost powers of the human brain; to build a better country-side we throw dice."

Land regulatory processes are often too narrowly focused, unevenly applied, and based on inadequate information. This promotes hostility among interest groups and leaves the general public with a sense of powerlessness and disenfranchisement. Most people are unaware or do not understand how land use decisions can dramatically affect their lives and neighborhoods.

Suburban jurisdictions often compete ferociously for business and development that once might have been located in the urban core. Municipalities lure businesses to their side of the border through tax breaks, infrastructure improvements, and other guarantees, but the costs of development, like increased congestion and pollution, are frequently borne by neighboring jurisdictions. With each county myopically focused on ways to increase its own tax base, the region as a whole becomes socially and economically fragmented. As jobs shift further from the central cities, people find they can live even further outside the metropolitan area and still have a reasonable commute to work. Those left behind in the older core cities, increasingly members of minority groups, face diminished job prospects, crumbling neighborhoods, and economic disparity.

The historical deference to local autonomy has, of necessity, precluded significant coordination among state and federal policies and actions. This disjointed approach has generated patchwork, *ad hoc* decisions. A basic challenge for land use policy in the future is to amend this approach to maximize environmental goals and reflect a broader sense of community.

Transportation and housing policies have been major contributors to wasteful land-use patterns. Transportation policies, designed almost exclusively for the automobile, greatly exacerbated suburban sprawl. Thousands of miles of trolley lines were abandoned or paved over to accommodate the car. The Interstate Highway Act of 1956 authorized construction of some 41,000 miles of new highways leading from cities to the hinterlands, and where the roads went, development followed. Business and suburban development flocked to the off-ramps of the new roads, but such growth came at the expense of cities and open space. The linkage between transportation and land use was rarely made, and national development patterns reflect that disconnect.

Federal housing policies also contributed to the growth of suburbia and the segregation of housing by class and race. In the decade following World War II, nearly half the houses built in the U.S. were financed with Federal Housing

Administration (FHA) and Veterans Administration assistance. These programs boosted a construction industry floundering after the Great Depression and improved the U.S. stock of housing. But FHA-backed mortgages were only available for new homes, primarily single-family, detached houses on inexpensive suburban land. The agency did not support loans to repair, remodel, or upgrade older houses in the cities that might have provided affordable housing for growing minority and immigrant populations. Cities reaped few of the benefits of the post-war development boom.

Poorly designed statutes, including some of the nation's environmental laws, have had unintended consequences. The Superfund program, designed to promote the cleanup of abandoned toxic waste sites, has failed to achieve its ends, despite its cost, and may actually hinder the reuse of abused lands. Even in cases where costs would be lower to recondition an old facility where infrastructure is already in place, lenders are reluctant to invest in such a project for fear of liability. The threat of liability for past contamination steers factories or urban renewal projects away from brownfields and encourages new development of "greenfields."

We are beginning to understand what we have lost and are unwilling to accept what has replaced this loss. Despite technological advances, we have produced housing developments that demean rather than inspire our citizenry. We have built mile after mile of ugly cookie-cutter houses, subdivisions devoid of character, congested streets, commercial strips that assault the eye with garish signs and neon lights all at the expense of townscapes, city cores, open space, productive farmland, and wildlife habitat. The costs of sprawl are not only aesthetic. The decline of cities and segregation of communities that results from land use decisions imposes measurable burdens on society. Local governments are increasingly aware that scattered large-lot zoning does little to protect habitat and often does not generate enough tax revenue to pay for municipal services. The environmental costs of poor land use practices are rarely factored into local decisions.

Growth is inevitable, but ugliness and environmental degradation are not. With forethought planners can channel growth to create more livable spaces and communities. Theodore Roosevelt called conservation a "great moral issue," and indeed our efforts to fashion a more sustainable society flow from a greater sense of reverence for the land and concern for present and future inhabitants. To pursue this ethic, we will need to identify more useful and understandable criteria for determining and measuring the costs of poor land uses. We will need to overhaul conflicting government policies that inhibit sound land use decisions. Land use planning depends on good information and the support of people at all levels of government, the private sector, and the citizenry to be successful. The following seven principles offer an approach to guide thinking about land use issues for the next generation.

Think Systems

Better land use planning can only be achieved if policymakers understand how development patterns impact natural systems. Long-term planning must consider systems landscapes, watersheds, estuaries, and bio-regions to be sustainable. Analyzing and abiding by the carrying capacities of systems must provide the basis for the development of our communities in the future. Since natural systems often cross political boundaries, cooperative efforts involving federal, state, and local entities, including businesses and private landowners, are critically important.

Tomorrow's professionals and decision makers will need to learn new tools and draw from multiple disciplines and then take the risk of working with experts from many fields. Transportation planners, educators, recreational experts, financial experts, health providers, and government officials must learn to come together and trade valuable information in a public format with farmers, businessmen, water quality specialists, wildlife biologists, and environmentalists. A much broader perspective is needed to assist communities to deal with the diverse and complex issues affecting their lives.

Water quality and quantity, for example, are closely tied to the use of land, and are of paramount importance to all people. Municipalities from New York to San Antonio are grappling with the need to protect open space and preserve water supplies in the face of increasing population pressures. But programs to protect and conserve water sources frequently extend far beyond city boundaries.

In a case that illustrates the need for systems planning and regional cooperation, state and local officials in New York have jointly developed a plan to manage growth and development in the Catskill watershed to preserve the water source for the 9 million residents of New York City. With foresight and financial commitments, city, state, and federal officials are putting together a solution for the residents of New York that protects a larger land area, provides needed fresh water, and saves hundreds of millions of dollars that would otherwise have to be spent on water treatment facilities for the city. Systems thinking requires a thorough understanding of the limits of the watershed and considers new development with that in mind.

Another example of the move toward a systems-based approach is the development of multispecies conservation plans to preserve threatened and endangered wildlife and plants. The Natural Communities Conservation Planning program in Southern California is an experimental effort to preserve the remaining coastal sagescrub habitat in an area of high land values and growth demands. The complex and often controversial plan impacts five counties and covers 6000 mi^2 and attempts to reconcile the conflicts between environment and development goals. Local, state, and federal partners are working cooperatively to carefully manage development, protect the threatened California gnatcatcher (*Polioptila californica*) and other imperiled species, and provide some long-term certainty for all stakeholders.

Community-Based Planning is Best

Sound land use planning requires local knowledge, involvement, and a community spirit to provide the energy, staying power, and creative ideas that can come when neighbor joins with neighbor in trust to mold a collective vision for the future. Fundamentally, land use planning is community based within a regional framework. Without the input and support of local people no plan can hope to succeed. Federal or state involvement may be crucial in providing overall guidance, startup technical assistance, baseline information, and funding resources to help communities and multiple local jurisdictions plan for the future.

While many people recoil from the thought of a federal land use policy, especially in the West, the reality is that the U.S. does have a policy. Transportation policies, farm programs, disaster relief, flood insurance, water and sewer support, wetlands, and endangered species laws, public housing, and financial lending programs combine to create a de facto national land use policy. An audit of federal programs affecting land use is long overdue to identify contradictions and move toward more consistent approaches to the use of land that complement regional and community goals.

Cooperation between governments is often difficult, but there are some models for integrating federal, state, and local needs. For decades, transportation infrastructure programs at the federal level were developed without regard to local or regional land use objectives. The Intermodal Surface Transportation Act (ISTEA) is a recent and innovative law that links transportation policy and investment with environmental concerns and local recreational needs, such as greenways and bike trails. Other models for cooperative land use planning at the federal level are the Coastal Zone Management Act (CZMA) and the Coastal Barriers Resources Act (CBRA). A voluntary program, CZMA provides federal assistance to states that develop coastal management plans and ensures that subsequent federal actions will be consistent with the plans. An innovative approach to encourage responsible land use planning, CBRA avoids regulatory mandates, but offers powerful disincentives by denying federal funds for roads, sewer plants, water systems, and flood insurance to developments that locate in sensitive coastal areas.

Fewer than a dozen states have comprehensive land use or growth management plans on the books, but those that do, like Vermont and Oregon, have realized impressive results. Florida, for instance, is experiencing explosive growth despite having management plans in place. Florida grew by an average of 892 people per day in 1996. Each day, 450 acres of forests are leveled, 328 acres of farmland are developed, and an additional 110,000 gallons of water are consumed. States can play a critical role setting ground rules for local governments and assisting municipalities in grappling with land use issues such as watershed protection that transcend jurisdictions. The ultimate objective of such plans is not to oppose growth, but to ensure that development is consistent with community and regional objectives. Environmental policies can be explicitly built into these plans, rather than

allowed to emerge incoherently as the function of thousands of disconnected land use decisions.

Perhaps the most significant achievements at the local level will come not
from government, but through the efforts of private citizens engaged in
place-based conservation. Born out of frustration with national organizations
or to promote a specific local issue, small grassroots conservation organizations have sprung up across the country. The proliferation of land trusts is
enlivening the conservation movement with new energy and excitement.
More than 1200 land trusts are now functioning across America, double the
number a decade ago and their numbers increase weekly. These diverse and
dynamic groups offer a fertile area for community ideas and involvement.

Better Information and Education

In deciding what kind of land use strategy to employ, a community must
understand its current makeup, strengths, limitations, and options. With the
information management technology of today, planners can review and
interpret seemingly infinite amounts of data on soils, vegetation, water
resources, biodiversity, view-sheds, tax structures, demographics, transportation and infrastructure needs, housing demands, recreation needs, and
other local priorities. These systems are enabling community planners to
develop models and make accurate predictions about the outcomes of policy
choices.

In Florida, for example, The Conservation Fund in partnership with the
MacArthur Foundation is using the technology of geographic information
systems (GIS) in a facilitation process that allows planners and citizens in
more than a dozen local jurisdictions to project possible growth management
options for the future of their region.

In northern Palm Beach and southern Martin counties, efforts are underway to reconnect the watershed of the Loxahatchee, the only federally designated Wild and Scenic River in Florida. Bringing together 18 different natural
resource public agencies addressing the watershed using GIS technology,
The Conservation Fund built a consensus on watershed restoration by creating a new interface for landscape and greenway planning called the Decision
Support Model. To build the human connections to nature, the project
focused on four greenway prototypes ranging from an historic, low-income
community to a new, neotraditional development. By connecting and protecting the green infrastructure of the region, and by building communities
that are compatible with the needs of the environment, we are not only ensuring the future health of the river, its watershed, and its wildlife, we are also
ensuring a sustainable future for the human communities of the region.

In an unprecedented effort in Alabama, seven major timber companies are
working with Auburn University and The Conservation Fund to gauge the
effect of different timber practices on an entire watershed and test timber
management strategies for their environmental impact.

Criteria must be developed for measuring the effects of land use decisions. Cost–benefit analysis can offer citizens and policymakers a better understanding of environmental and economic costs of land use. Quantifying the overall costs of sprawl would help communities assess how best to manage growth in their region.

For communities to take a lead in promoting sound land use policies, individual citizens will need a better understanding of the impact of land use choices on the environment and their future quality of life. Significant change will not soon occur in land use planning unless the public demands it. The more people understand these issues, the more likely a constituency will emerge for good land use planning. In short, we need to increase the ecological literacy of our citizenry. Ecological education at all levels should provide information about the relationship between the human environment and natural systems. Citizens must understand the inherent links of land use with clean air and water, safe and healthy neighborhoods, a prosperous economy, and a stable tax base if they are to be empowered to take action.

Many studies have detailed the high costs of suburban sprawl for municipal governments that are hard-pressed to pay for police and fire protection, schools, water systems, and sewers. A recent study done by Culpeper, VA, found that for every $1.00 in tax revenues from residential development, the city must pay $1.25 to provide necessary services. The same study, conducted by the Piedmont Environmental Council, found that for every $1.00 in taxes collected from farms, forests, open space, or commercial lands, 19 cents was paid out for services. Large-lot exclusionary zoning can be costly, but many planners and citizens still cling to the notion that such practices are inherently profitable.

With information and education, communities can begin to develop the vision and leadership to build a more sustainable future.

Build Partnerships

Land use decisions are often controversial, but a growing number of enlightened leaders from various perspectives now recognize how much more can be accomplished when ideologies are checked at the door and rational people sit down to discuss solutions. Government, industry, nonprofit organizations, and citizens can have much greater impact working together than any one of them could have working alone. Next-generation policies must include new models of collaboration to avoid the rancor of our traditional adversarial approach to environmental issues.

Nowhere has there been more acrimony than the debate over endangered species protection. Increasingly, however, private land owners, corporations, and the federal government are coming together to form habitat conservation agreements to protect imperiled species. These agreements provide certainty to landowners while ensuring an adequate level of protection for the affected species. In another example, the governor of Maine, environmental organizations, and timber companies in Maine met to write a compact limiting clear-cutting and improving forest practices across the state. The compact was placed on the ballot in the 1996 election as an alternative to a more extreme measure. The compact passed by a wide margin. Such initiatives were unheard of just a decade ago.

An excellent example of partnerships is emerging in the Sustainable Everglades Initiative (SEI). The Everglades, a unique and diverse biosphere in North America, is threatened by human-altered hydrologic processes. The SEI brings together public and private stakeholders to develop comprehensive, whole-system approaches to sustainability in the Everglades and South Florida. SEI is a learning strategy focused on rethinking what it means to be "citizens" of South Florida. SEI participants bring together diverse issues and perspectives, creating integrated economic development, community development, and environmental restoration strategies around an evolving ethic of sustainability. Working with the Florida Department of Community Affairs Eastward-Ho! Initiative, participants of the projects are developing creative and collaborative strategies for redevelopment in the urban communities of southeast Florida and the Glades communities near Lake Okeechobee.

In the next century, significant gains in environmental quality will be the result of private-sector initiatives. Business leaders whose expertise, experience, political savvy, employees, and resources must be engaged to address thorny issues like one-point source pollution and biodiversity protection. Private landowners now hold most of the remaining wetlands, endangered species habitat, timberlands, and open space in the nation. Partnerships between public officials, private groups, and major timber companies are already providing ways to harvest timber while expanding outdoor recreational facilities, restoring streams, and restoring habitat for threatened species.

We will also find ways to engage more private landowners in conservation. The "Partners in Wildlife" program of the U.S. Fish and Wildlife Service, for example, provides funding and technical assistance to 25,000 farmers and ranchers who want to restore wildlife habitat on their lands. Landowners get the satisfaction of improving the environmental quality of their property with assurance from the federal government that new regulations will not be imposed on their property should they choose to return the land to agriculture.

Empower the Disenfranchised

Resource planners and environmentalists generally have failed to reach out to the diverse social and economic groups in America. Often environmental quality is seen as an elitist issue of little concern to the poor. Many environmental activists have been slow to make the connection between declining cities and the loss of open space, between social issues and natural resource issues. Conservation should not be about preserving special places for the wealthy; it should be about improving the quality of life for all our citizens. We will never achieve success in conserving natural habitats if we ignore the human habitats crumbling in our midst. Poverty, joblessness, and unsafe streets are environmental problems as well.

The rise of the environmental justice movement has led to much broader participation. From clean water, to lead paint, to brownfield redevelopment, environmental concerns affect all Americans regardless of race or socioeconomic position. The collaboration of people interested in environmental, social, and inner-city concerns will help change the way we think about land use issues. In the future, conservation, transportation, and development policies will thus take into account less affluent and less politically powerful members of society and galvanize inner-city groups to become active on environmental issues in their communities.

Nowhere is the cumulative effect of land use decisions more evident than in our cities. For example, because of a fear of crime and loitering, playgrounds, basketball courts, and community centers are often neglected or never built. Charles Jordan, an African–American leader and director of Parks and Recreation in Portland, OR, observed, "We are the first generation in history that fears its children. This fear can have a spiraling effect less positive recreational opportunities, more antisocial behavior, more fear, and ratcheting down of the services we offer." Private citizens, church, and civic leaders must begin to act on multiple fronts to counter the continued social stratification and decay of its cities and urban people. The decline of cities as well as rural areas is everyone's problem and therefore everyone needs to be part of the solution.

Protect and Enhance Wildness

Wildness is not a faraway place, but a spirit, a characteristic of complex natural systems and places. Wildness might be found in a small wood lot, native grassland, in a pond with tadpoles, or in a backyard visited by migratory songbirds. Wildness speaks of beauty, resilience, diversity, challenge, and freedom. Wildness is one quality that defines us as a nation and uplifts us as a people. In protecting wildness, we are protecting something in ourselves. We sustain not just the tangible benefits of natural systems, new medicines, genetic materials for crops, air and water quality, but the character and staying power of the earth itself.

Protecting wildness as a national policy was an American invention. We were the first country to establish national parks, protect forests, establish wildlife refuges, scenic rivers, and first to protect endangered species. But there remains a need for more open and natural spaces in highly populated areas where little public land exists and outdoor recreational opportunities are few. A 1995 study conducted for a group of the largest home builders in the nation found that Americans increasingly want to be able to interact with the outdoor environment in the places they live, through trails, prairies, woods, and open space. In that survey, 77% of respondents selected "natural open space" as the feature they would most like to see in a new home development.

Florida, Maryland, and other states have begun ambitious programs aimed at establishing greenways and protecting open space corridors for wildlife and recreation. More needs to be done, and hopefully there is still time. While voters repeatedly support bond issues to fund land acquisition, the lack of a coordinated constituency for public lands has allowed Congress to divert more and more acquisition funds away from the Land and Water Conservation Fund, an account established to meet federal, state, and local public land needs. Acquisition funds should be restored and new sources found. The rise of the land trust movement will reap tremendous benefits for open space while relieving public maintenance and acquisition burdens. These citizen-driven efforts offer one of the best hopes for conservation in the next century. We must experiment with new collaborative approaches like scenic easements, tax credits, transferable development rights, reducing estate taxes, and technical assistance to encourage the retention and restoration of as much wildness as possible.

Renew Spirituality

Conservation is sometimes difficult because it is in many respects a moral issue. It requires a sense of values, caring, and charity, a reverence for the blessings of nature and a shared commitment to the stewardship of the earth. We need not repeat the mistakes of this century in the next, but assuredly we will if we fail to take stock of our actions and accept responsibility for our land use choices. All our natural resource "stuff" has simply become too secular. A renewed sense of morality and passion must be instilled in how we use the land and its products, how we care for one another, and what kind of places we leave for future generations.

The best-selling author and philosopher Thomas Moore observed, "The greatest malady of the 20th century is the loss of soul." There is in America a growing disquiet, noted by many commentators, liberal and conservative alike, that the nation is losing its sense of purpose and morality. Loss of community, the "breakdown of society," is an oft-heard refrain. Our efforts to foster land stewardship and connect people to the land is an attempt to focus individual attention on common goals and values. While people may differ in their religious, cultural, and ethical ideas, a sense of respect for nature is common to many traditions. It is time for a rediscovery of what we believe to be right and wrong.

Changing the relationship of people to the land will not be easy. American laws governing land use have always been based on the premise that land is a commodity to be bought and sold for capital gain. Aldo Leopold put it most eloquently: "We abuse the land because we regard it as a commodity belonging to us. When we see land as a commodity to which we belong, we may begin to use it with love and respect. There is no other way for land to survive the impact of mechanized man, nor for us to reap from it the aesthetic harvest it is capable, under science, of contributing to culture. That land is a community is the basic concept of ecology, but that land is to be loved and respected is an extension of ethics. That land yields a cultural harvest is a fact long known but latterly often forgotten."

Conclusion

Forging an ethical relationship with the land and its people is the challenge of our time. These seven principles offer only a guide for creating new tools and methods of decision making that will shape the character of our national heritage. Improved land use policies will need to be based on a systems approach that reduces the waste of land and resources, enhances wildness and community character, permits growth and economic development, and preserves healthy and functioning ecosystems. No net loss of greenway should be our goal for the 21st century. We must find ways to accommodate projected growth while preserving open space, farmland, watersheds, and rural communities. Redevelopment of brownfields and abandoned property must be afforded a higher priority than development of virgin lands on the metropolitan fringe.

We will need to develop more balanced, fair, and flexible regulatory approaches and reexamine government programs and procedures at all levels. Initiatives by local interests, public and private, must be encouraged, but additional leadership needs to come from state and federal governments that can better coordinate actions that promote regional growth management objectives. Given that public support is the requisite for progress in the land use arena, we must make sure that local and statewide constituencies are developed, nurtured, and strengthened. More multidisciplinary approaches to training professionals must be developed. Increased education and outreach to new partners is critical to success.

Land use planning was about people deciding what their communities should look like in the future. It is not a radical idea, but it will require leadership, vision, and innovation. The payoff is a better quality of life, a stronger economy, and a healthy environment for the future of America.

Acknowledgment

This presentation was prepared in part with The Next Generation Project, Yale Center for Environmental Law and Policy, Yale University.

10

The European Experience: From Site Protection to Ecological Networks

Rob H. G. Jongman and Daniel Smith

CONTENTS

Introduction

Ecological networks are the result of science-based nature conservation. Its basis is founded in biogeography, population dynamics, landscape ecology, and land use science. That means that they do not only consist of ecological elements, but also political, planning, land use, and awareness components. Without incorporation of these aspects ecological networks cannot be realized.

1-56670-368-9/00/$0.00+$.50
© 2000 by CRC Press LLC

Ecological knowledge on ecological networks is based on insights in landscape hierarchy (O'Neill et al. 1989), biogeography, population dynamics, and landscape change. Landscape hierarchy is the first basic principle to classify the levels and the systems of ecological networks. Corridors and sites of importance on the continental level differ from those on the regional, state, or country level. Biogeography is important to define the role of species and the national or international responsibility for a region or country. For instance, the European beech (*Fagus sylvatica*) is common in all Europe, but absent outside. It is a European task to maintain its natural area and habitat diversity. The Pyrenean oak (*Quercus pyrenaica*) occurs only in southwestern France, northwestern Spain, and northern Portugal. Its protection is a task of these three countries. Spain and Portugal together have the task to conserve the Iberian lynx (*Lynx pardina*).

There are more than 50 countries within Europe, and each has a different phase of policy development, a different planning system, and awareness of nature conservation as expressed in development of nongovernmental organizations differs greatly among them. The differences depend on the history of the countries, both in economic sense and in political sense. It is obvious that countries in Central and Eastern Europe have developed differently than the countries in Western Europe. But there is also a difference between the northern and southern countries, mainly based on the development of democratic structures which influenced nature conservation strongly through development of awareness (and social influence on political decisions), economics and the possibility to found organizations.

Not only man travels and makes use of roads. Natural species can also migrate over long distances and they also move through the landscape in search of food, shelter, and new breeding sites. They travel at different scale levels, constructing their own pathways and their own network.

Migrating species are especially vulnerable. They cannot at every moment be identified as being present and they often compete with human land use. European storks (*Ciconia ciconia*) for instance, breed in northern Europe and winter in Africa, migrating 10,000 km each season. The breeding population is mainly concentrated in Germany, Poland, Czech Republic, the Slovak Republic, Hungary, and the Baltic states in the east and Spain and Portugal in the west. They used to cover a larger area, but their breeding success was severely hampered by land use changes in the last decades. In the Netherlands for instance, where several cities have the stork in their heraldic weapon (similar to a family's coat of arms), in 1996 only two pairs of wild storks are left.

What is the stork habitat? It consists of wetlands and especially grassy wetlands where large insects and other small animals like mice and frogs can be found. Drainage and agricultural intensification changed their foraging habitat and in this way nearly caused its extinction, as happened to the otter (*Lutra lutra*) in the Netherlands as well.

Land use change in Europe happens through ages. Europe consists of rather restricted areas of natural landscapes and large areas of cultural landscapes, made by man and showing the diversity of the regional climate and soils. Through centuries this has led to a pattern of landscapes that was rather stable until the second half of the 19th century. Then the industrial revolution took place. It meant not only a revolution in the urban environment, but also in the rural environment. Machines were introduced, as well as fertilizer and wire fencing. This meant that seminatural areas were converted into agricultural land and that the scale of agricultural holdings was increasing. In the same time the main European rivers started to be regulated, parts of the Rhine, the Danube, the Elbe, the Meuse, and the Tisza. That meant better transport facilities, less fish migration, and better drainage. This process started on a small scale of course, but continued until now. Changing our environment has been one of the major issues for the deterioration of nature. It caused:

- increasing land use intensity;
- larger units both in nature and in agriculture;
- sharper boundaries between nature and agriculture; and
- both population enlargement and fragmentation of natural populations.

Species have adapted to the cultural landscapes of Europe, because they were accessible and not hostile and because of the small-scale character. However, it seems that the changes ongoing since the last decades will lead to the extinction of many species unless habitat quality improves and the landscape structure is restored.

Nature Conservation Development in Europe, from Action to Planning

Nature conservation in Europe has been inferred from developments in society, although with differing speeds in different countries. In the beginning of this century it was a reaction of scientists, teachers, artists, architects, and other educated people against the destruction of nature by the industrial revolution. The technological and economic development lead to an increasing loss of nature. At the same time the valuation of the beauty of nature, the love for nature, and the recognition of its importance for outdoor recreation increased and this was expressed in literature, art, architecture, and urban planning. The controversy between valuation of nature and the loss of nature created a basis for the beginning of nature conservation. In this first period nature conservation was based on private initiatives organized through the foundation of voluntary organizations. In many parts of Europe this moment can be located at the turn of the 19th to the 20th centuries (Bischoff and Jongman 1993). In this period three types of organizations based on different visions on nature conservation can be distinguished:

1. organizations following the ideas behind the foundation of the national parks in the U.S.A. (Yellowstone, 1872);

2. organizations aiming at the conservation of the values of naturde-scribed by scientists and artists like Alexander von Humboldt (Germany) and Jean Lahor (France); and

3. organizations emphasizing the importance of bird protection for human uses.

The first group was focused on the foundation of national or nature parks according to the examples in the U.S. Central is the conservation and, if necessary, restoration of natural and seminatural values in large areas. In the countries where these organizations had influence national parks and nature parks have been developed which cover rather large areas.

The main activities of the second group were the protection of areas with high natural values in combination with historical landscapes, often in relatively small nature reserves or as extensively managed historical landscapes. Some of these organizations used the strategy to buy the most threatened areas and to manage them. This is practiced by such organizations as the Vereniging tot Behoud van Natuurmonumenten (The Netherlands), National Trust (United Kingdom), Natuurfredningsforening (Denmark), La Ligue Luxembourgoise pour la protection de la Nature et de l'Environment (Luxembourg), Ligue Belge pour la protection de la Nature (Belgium).

The third group focused on bird protection. These organizations also acquired reserves and reached a high degree of acceptance by the people and the governments. In all cases these organizations were rather effective in realizing legislation for bird protection. Examples are Ligue française pour la protection des Oiseaux (France), Deutscher Verein zum Schutz der Vogelwelt (Germany), Nederlandse Vereniging tot Bescherming van Vogels (The Netherlands), The Royal Society for the Protection of Birds (United Kingdom), and Ligue Luxembourgoise pour la protection des Oiseaux (Luxembourg).

EU Habitats Directive and Ecological Networks

The European Union adopted in 1992 the Habitats and Species Directive (EC 92/34), meant for the conservation of natural habitats and species. The core of the Habitats Directive is the development of "Natura 2000," a network of special areas for conservation (SACs). In article 10 it is stated that national or regional governments can develop a policy to support "favorable conservation status" in the core areas. Core areas and the species in them can be supported by measures in the wider landscape. The Habitats Directive indicates that SAC are sites of community importance designated by the member states through a statutory, administrative, and/or contractual act where the necessary conservation measures are applied for the maintenance or restoration, at a favorable conservation status of the natural habitats an/or the populations of the species for which the site is designated.

The conservation status of a natural habitat is favorable when:

- its natural range and the area it covers within that range are stable or increasing,
- the specific structure and functions which are necessary for its long-term maintenance exist and are likely to continue to exist for the foreseeable future, and
- the conservation status of its typical species is favorable.

The conservation status of a species is favorable when:

- population dynamics data on the species concerned indicate that it is maintaining itself on a long-term basis as a viable component of its natural habitats,
- the natural range of the species is neither being reduced nor likely to be reduced in the foreseeable future, and
- there is, and will probably continue to be, a sufficiently large habitat to maintain its populations on a long-term basis.

Spatial transition from one biological community to another has attracted the interest of ecologists, geographers and wildlife and land managers for several decades. "Ecotones," "buffer zones," and "natural corridors" (and related or synonymous concepts) are concepts relying on the idea of transitional zones between ecological units. These concepts for nature conservation have recently been enriched by recognizing their value regarding biodiversity maintenance and control of flows across the landscape. A landscape is a network of patches or habitats connected by fluxes of air, water, energy, nutrients, and organisms. Interactions between habitats are thus defined by these landscape fluxes and the function of the latter for certain habitat conditions.

If an area is a SAC for Natura 2000, being a representative sample of the biodiversity of Europe, however, does not mean that it stands alone. It should function as an optimal habitat for the species concerned and function without disturbances from the outside. They should even function for the wider environment as a source and refuge area for species. That means that linkage with the wider landscape is essential. This also means a link with policies for the wider countryside; policy and planning for the supporting areas mean also linkage between nature conservation, agriculture, and the realization of road and railway networks. Here integration between national and European policies is vital.

Buffer zones and ecological corridors are management objects which may be necessary to ensure the conservation status of species and habitats within the Natura 2000 sites. There is a need to consider features required across areas, and set out the overall character of an area which is necessary to achieve a favorable conservation status. This will include consideration of the full range of ecological needs of the species involved, including movement, dispersal, migration, and genetic exchange.

The Habitats Directive refers to corridors and stepping stones. We need to be neutral as to shape and extent of corridors: one important contribution they can make is to ensure a sufficient habitat to maintain populations across their total natural range. This will require decisions on location, management, and pattern. This is clearly flagged in the Birds Directive (article 3(2)(b), (c), and (d)) and is part of the Habitats Directive. Article 10 states that the responsible authorities can take measures in the wider landscape to enforce the favorable conservation status and the functioning of SACs by protecting or managing linear features such as rivers, streams, and hedgerows. It has been identified as a national or regional responsibility to decide on that.

All kind of linear elements on different scales, such as single hedgerows, small streams at the lowest level and hedgerow landscapes, patchy forest landscapes, and rivers on an intermediate to continental level can fulfill this function. Hedgerows and first- and second-order streams are key elements on the local scale. They provide food, guidance, and shelter for small mammals, birds, and amphibians; they also are the wintering sites, nesting sites, spawning grounds for fish species, and the transport route for river-transported plant species. Larger rivers and related wetlands can provide foraging grounds for large mammal species, migrating birds, and river fish on a larger, even continental scale.

After the Second World War, nature conservation was focused on the preservation of values within seminatural landscapes. This was especially important in the northern states of Europe, where the decline of nature was alarming. In the 1970s, many changes took place in nature conservation; nature conservation acts were revisited in several countries. Some countries amended the existing legislation, others formulated a wider nature conservation policy and included relations with other policy issues (recreation, urbanization, regional planning, and agriculture). This period can be characterized as the time of acceptation of responsibility of nature conservation by national governments.

In all parts of Europe, landscape ecology as a science evolved from the 1950s on. There has been exchange between Western and Eastern Europe, but the great difference was in the influence that science could have on planning and policy. Introduction of landscape ecological principles by Troll in the 1950s and later by Zonneveld in the 1960s forced ecologists to look outside their laboratories and outside their protected areas. Hierarchy in landscapes, flow principles, time–space relationships, and, later, island biogeography theory and metapopulation models made nature conservation organizations doubt on their long-term success, especially because of the many small nature reserves and the breakdown of the connectedness of the landscapes of Europe. After the first European nature conservation year 1970 planners at regional and national levels were asking landscape ecologists how to deal with nature in spatial and regional planning. Scientists were forced to think about nature in a holistic way and to discuss their results on costs and effectiveness with other parties in society.

In the last decades of the 20th century, nature conservation strategy changes strongly and starts to adopt landscape ecological principles. This change takes shape in new strategies formulated in policy documents on nature conservation and nature rehabilitation of former or potential natural areas. This stage in the development of nature conservation can be seen as a period of cooperation worldwide and within the European Community (EC, later European Union, EU). Development of nature conservation is occuring at least as far as it concerns organization and legislation.

The acceptation of the Bern Convention calls upon the contracting parties to take action to maintain wildlife populations, to develop national policies on wildlife conservation, and to control pollution and other threats to wild flora and fauna. This lists endangered migratory species of mammals, birds, reptiles, fish, and insects and obliges member states to take steps to protect listed species and control pressures upon them. This convention as well as the Bonn convention on migrating species has been supported by the EU and has even translated into EU legislation. The EC directive for the Conservation of Wild Birds (EC/79/407), agreed on in 1979, emphasized the need for international action on bird protection and set out provisions for the protection, management, and control of all species of naturally occurring birds in their wild state in the Community territory. It is the translation of the Bonn convention into EU legislation. In 1988 the first EC proposal was made on a directive on the conservation/protection of natural and seminatural habitats and their wild flora and fauna. It was agreed on in 1991 and came into force in 1992.

National differences are expressed in legislation and planning and on the definitions used in it. This means that in Europe the differences between national nature conservation policies will be seen in the definitions of, for instance, national parks. For international planning, understanding each other and knowing differences in definitions is a prime issue.

In policy development for nature conservation in Europe the national level and the regional level of planning are important. At these levels ecological networks emerged here as realistic principles since the 1980s in both Western and Eastern Europe. Coherent European approaches are relatively young. This means that there is a whole diversity in approaches in Europe, hopefully with the same objective. The recent developments tend to international coordination. The convention of Bern on protection of European wildlife was the first attempt in that direction. The EU translated that initiative into its Habitats Directive. The Habitats Directive includes more or less a principle of an ecological network, although it leaves much of its realization to national governments. The convention on Biodiversity has been signed and ratified by most European countries. It has been decided not only to make national biodiversity strategies, but also a Pan-European approach, the Pan-European

Biological and Landscape Diversity Strategy (Council of Europe, 1996) of which the Pan-European Ecological network is the core element, linking and coordinating all national initiatives.

Definitions of National Parks in Europe

IUCN:

A National Park is a relatively large area, where

- one or more ecosystems have not been changed fundamentally by human exploitation and habitation, where plant and animal species, geomorphologic objects, and biotopes of special value occur or that contain a natural landscape of great beauty;
- the highest authority in charge of the country took steps to avoid potential exploitation as soon as possible, to reduce settlement in the whole area, and to stimulate effectively the conservation of ecological, geomorphologic, and aesthetic characteristics that led to the initiative of its foundation; and
- it is allowed to visit the area under special conditions for the inspiring educational, cultural, and natural values.

The area must be managed as a whole.

Germany:

A National Park is an area of a larger size, slightly influenced by man and that deserves special protection because of its natural beauty and special ecosystems and where the core area is managed as a nature reserve.

The Netherlands:

A National Park is a single area of at least 1000 ha consisting of natural systems such as waters and forests with a special condition and plant and animal live. Good possibilities exist for zoning and recreational use as well. In a National Park, nearly no agricultural land is found.

Great Britain:

A National Park is a large area mainly founded because of its great landscape and scenery values. Human settlement and human activities are usually present. The importance of the natural environment varies per park and, if present, natural values are situated in nature reserves in the park.

Greece:

A National Park is an area that is mainly forested and needs special protection because of:

- flora, fauna, geomorphology, soil, air, waters, and natural environment in general,
- the necessity to keep the natural condition undisturbed or to improve it because of aesthetic values, welfare of man, and scientific research.

Italy:

A National Park is a large area

- that is protected because of the presence of valuable flora and fauna, important geological formations and landscape beauty,
- that aims at the enhancement of recreation and tourism,
- that gives space to human exploitation to provide an income for local people, and
- where it is forbidden to hunt.

Portugal:

A National Park is a large area that can be found in remote parts of the country where man manages the environment in the same traditional way he has for centuries. A National Park contains special landscapes and an important flora and fauna.

France:

A National Park is an area that is nearly uninhabited, with strict rules for conservation of flora and fauna, biotopes, and special landscapes for visitors of the Park and that is surrounded by a buffer zone in which tourist activities and rural economy will be stimulated.

The Scientific Basis of Ecological Networks

When thinking of the realization of ecological networks, it is not only the national, but also the regional and even the local level that are of importance: on the latter decisions have to be taken on what, where and how small sites and corridors will be realized. There you have to decide where and how to allocate sites and corridors and eventually where to rehabilitate nature. You need to base that on ecological data and land use data. Most models are based on rules of thumb. However, it is possible to develop allocation models that can be evaluated with succession and metapopulation models. For several areas and species in the Netherlands such allocation models have been developed based on suitability for agriculture and nature conservation (Reijnen et al. 1995). Differences in suitability make it possible to design the best possible pathway and calculate its ecological effectiveness and its costs.

Landscape ecology and, embedded in it, population dynamics, give a scientific basis to nature conservation strategies. They provide the insight that nature is a relatively dynamic system reacting on a complex of environmental and land use conditions. Land use is considered to influence the functioning of ecosystems as a whole, its self-purification capacity and the carrying capacity of the landscape (Mander et al. 1988; Kavaliauskas 1995). It also affects habitat quality for wild species and the potential for dispersal that is vital for survival of populations especially in fragmented landscapes.

Large areas with good living conditions that are always inhabited are defined as core areas for populations. In good reproductive years species will move from these areas into other marginal sites (Verboom et al. 1991). Area reduction will cause a reduction of the populations that can survive and in this way an increased risk of extinction, because dispersal between habitats decreases, causing less exchange of genetic information and a reduction of the colonization of empty habitats.

Most natural and seminatural habitat sites are remnants of a former natural area. In the time that Europe was covered merely by natural and seminatural vegetation, species within these forests and scrubs—in general the less dynamic habitats—had no problems of dispersal or migration. Their biotopes were large and well accessible. Dynamic ecosystems were present as well, but were relatively small, and species were adapted to quickly disperse and colonize the biotopes. However, it appears that even in production forests management can cause isolation of the remnants of natural old-growth forests within it (Harris 1984). Nowadays isolation is an important feature in agricultural landscapes of Europe.

Plants and animals both disperse by wind, water, with help of other species, or by their own movements. Migration is a specification of dispersal, while it is directed to a certain site. Dispersal is essential in population survival and the functioning of biotopes. However, dispersal can only function if there are sites to disperse from and to and means for dispersal. Dispersal is important for survival of populations. On the one hand animal species will leave a population if living conditions cannot support all individuals; on the other hand species will fill in gaps in populations or sites that become empty. Fluctuations in populations can cause changes in species abundance and species composition of a site. Birth, death, immigration, and emigration are the main processes regulating fluctuations at the population level. Plants, and several other groups of species depend on other species for their dispersal. Restriction of species dispersal increases the chance of species extinction (den Boer 1990).

The main elements in the landscape of importance for dispersal are the distance between sites, the presence of corridors, and the barrier effect of landscape and land use between (Opdam 1991). Area reduction will cause a reduction of the populations that can survive and in this way create an increased risk of extinction. It also will increase the need for species to disperse between sites through a more or less hostile landscape. Routes for species migration consist of zones that are accessible for the species to move from one site to another and back. Migration routes can be manifold, from single wooded banks to small-scale landscapes and from river shores to whole rivers and coastlines. Migration is a prerequisite for many species from northern Europe to survive the winter period. For flying animals this means that their route must lack barriers and that stepping stones must be available for feeding, rest, and shelter. For fish it means that rivers are not blocked by dams and that they are of good water quality. For mammals and amphibians it means that guiding greenways are available and that man-made barriers can be crossed. If the dispersal between habitats decreases, isolation will cause less exchange of genetic information and a reduction of the colonization of empty habitats.

Ecological Networks in the Pan European Biological and Landscape Diversity Strategy

At the conference of the European Ministers of the Environment in Sofia on October 25, 1995, a declaration was adopted in which the Ministers stated, among others, that

> Recognising the uniqueness of landscapes, ecosystems and species, which include, inter alia, economic, cultural and inherent values, we call for a Pan-European approach to the conservation and sustainable use of shared natural resources. We endorse the Pan-European Biological and Landscape Diversity Strategy, as transmitted by the Committee of Ministers of the Council of Europe for adoption at this Conference, as a framework for the conservation of biological and landscape diversity. We welcome the readiness of the Council of Europe and UNEP, in co-operation with OECD and IUCN, to establish a Task Force or other appropriate mechanism in order to guide and co-ordinate the implementation and the further development of the Strategy. In this respect we request the widest possible consultation and collaboration in order to achieve its objectives with a view to reporting on progress at the next Conference.

The strategy is prepared with the aim of supporting European implementation of the World Conservation Strategy, Agenda 21 and the Convention on Biological Diversity, the European Conservation Strategy and Helsinki Summit Declaration, Bern Convention, Bonn Convention, and EU mechanisms principally under the 5th Environment Action Programme and Natura 2000. It has a working period of 20 years.

The operational framework is based on sustainability and integration of biological and landscape diversity into all economic and social sectors. The strategy aims to introduce sustainable management to viable areas of biological diversity, and to introduce ecological network elements such as corridors, buffer zones, and stepping stones to increase viability of smaller areas. The Strategy is worked out into an Action Plan that is the basis for the implementation of the short-term goals of the Strategy with actions for a five-year period. In the next 20 years, the Strategy seeks to introduce biological and landscape diversity considerations into all social and economic sectors by striving to integrate them into agriculture, forestry, hunting, fisheries, water management, energy and industry, transportation, tourism and recreation, defence, structural and regional policies, and urban and rural planning. Main actors that would be involved in the implementation of the Strategy would include national authorities, bilateral donors, international organizations and financial institutions, organizations and associations active in the economic sector, private enterprises, the research community, information dissemination organizations, private and public landowners, nongovernmental organizations, the public (grassroots and citizen groups), and indigenous and local peoples of the regions of Europe.

The action themes can be divided into three groups, (a) organizational, (b) integrative actions, and (c) ecosystem and species-oriented actions. These are

a. Organization oriented: (1) Pan-European action to set up the Strategy process.

b. Integrative: (2) establishing the Pan-European Ecological Network, (3) integration of biological and landscape diversity considerations into sectors, (4) raising awareness and support with policy makers and the public, and (5) conservation of landscapes.

c. Ecosystem and species oriented: (6) coastal and marine ecosystems, (7) river ecosystems and related wetlands, (8) inland wetland ecosystems, (9) grassland ecosystems, (10) forest ecosystems, (11) mountain ecosystems, and (12) action for threatened species.

The development of the Pan-European Ecological Network is the key action theme of the strategy. Priority actions are designed to ensure that the Pan-European Ecological Network can be implemented within ten years:

• Establish a development program for the Pan-European Ecological Network. The development program for the Pan-European Ecological Network will design the physical network of core areas, corridors, restoration areas and buffer zones.

• Develop the first phase of an implementation program for the Pan-European Ecological Network. The development program for the Pan-European Ecological Network will be supported by the preparation of an implementation program. The implementation program will set out the actions that will be necessary to ensure that the Pan-European Ecological Network is created by 2005.

- Stimulate the development of national ecological networks and their linkage with the Pan-European Ecological Network. Ecological networks are being developed in a large number of European countries. These networks can make an important contribution to both the design of the Pan-European Ecological Network and its implementation at the national and regional level.

- Promote awareness of the Pan-European Ecological Network. Provide opportunities for exchange of expertise between countries in Europe on effective education and communication policies, with emphasis on the Pan-European Ecological Net-work, national ecological networks, and the integration policies.

Spatial scale can differ from local to continental and global. As the distance between suitable biotope sites increases, the number of species that can bridge this distance decreases. Ecological corridors and stepping stones can be essential for long-term persistence of species. Ecological relations are found to be of all kinds, through air, in the water, and on the ground. Ecological corridors can be of all kinds as well, and that makes it difficult to define them and to realize them in practice. Now, research is carried out and by several groups, so it might be expected that criteria can be set in the near future. However, the species approach might not be the only solution because of its restricted potential for generalization. Also, research on landscape functioning should be carried out to analyze the functions of landscape structures and landscape processes.

Amphibians and mammals are able to disperse over distances from several meters to hundreds of kilometers. For small mammals ecological corridors can be hedgerows, brooks, and all kind of other natural features that offer shelter. For forest birds small-scale landscapes characterized by a certain density in wooded banks can function as corridors from one forest to another. Birds like geese use northern Europe for breeding and southern Europe for wintering. Migration is important for grazing animals like deer (*Cervus elapus*) and roe deer (*Capreolus capreolus*), for predators like the golden eagle (*Aquila chrysaetos*), the pardel lynx (*L. pardina*), and the wolf (*Canis lupus*), but also for most birds from northern and eastern Europe. Other birds, swallows and storks use the European continent as part of their migration route to Africa. Salmons and sturgeons move up the rivers. Migratory species are not only depending on their breeding habitats but also on the presence of temporary habitats, within their migration route. These stepping stones are used for feeding and resting during migration as is the Waddenzee for many Fennoscandian species.

Not only fauna, but also flora can move from one site to another with help of wind, water, or animals. It is a means for dispersal, to colonize new sites, or to escape changing environmental conditions. However, plant strategies for dispersal are the least known and are difficult to detect in practice.

The Structure of Ecological Networks

An ecological network is composed of core areas (usually protected by) buffer zones, and (connected through) ecological corridors (Bischoff and Jongman 1993). Ecological corridors and buffer zones have already a history of scientific investigation and application and are becoming key elements of the "ecological network" strategy. Ecological networks are more widespread in Europe than many would suppose (Jongman 1995). Reviewing recent developments concerning ecological networks, Arts et al. (1995) concluded that "during the last decade, the nature conservation policies in many European countries have been based on landscape-ecological research, especially concerning the role of land use and landscape structure in the survival of species and in the protection of nature reserves. Plan proposals were made to establish ecological networks on local, regional and national scales."

Habitat quality is the major factor in determining if species occur in an area. Availability of food, shelter, and breeding condition determine if species can survive. The occurrence of wildlife species within a certain habitat is also

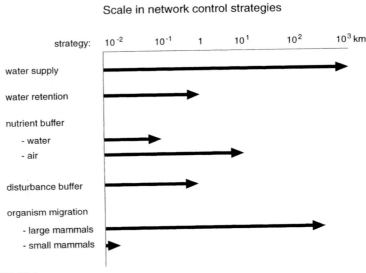

FIGURE 10.1
External conditions for habitat functioning consisting of three groups: (1) water supply and retention, (2) buffering impacts, and (3) supporting dispersal. (Farjon, in Jongman and Troumbis 1995. With permission.)

determined by three external habitat conditions: water supply, buffering of nutrients, energy and human impact, and dispersal of species (Figure 10.1). Within a habitat network the optimal habitat condition is related to a certain steady state of landscape fluxes, such as air movements, water flows, species migration, and human transport. Within the perspective of landscape fluxes suboptimal habitat conditions in a designated area can result from too small or too large inputs and outputs of water, matter, energy, organisms and human influence. Buffer zones and ecological corridors often coincide with multifunctional landscapes. These landscapes support both nature and other land use functions. Buffer zones and ecological corridors are nature conservation specifications of these multifunctional landscapes. By changing human influence in these parts of the habitat network habitat conditions in the core area may regain their former optimal status. That means that by managing buffer zones and ecological corridors the favorable conservation status of habitats and species can be supported. In practice landscape linkage elements, especially ecological corridors, are not yet integral parts of conservation plans. Implementation of theory and research results in policy and management practice is a long-term process (Jongman 1995).

Buffer Zones

The concept of buffer zones is rather old (Wright and Thompson 1935). Literature offers a number of definitions, related to the approach used for their design within the framework of a spatial and management strategy. Within a hardly anthropocentric view of nature, Jehoram (1993) distinguishes the nature management approach and the landscape approach generating different definitions, allocation criteria, and management strategies for buffer zones. The nature management approach can be illustrated by The World Conservation Union IUCN definition of buffer zones: a zone peripheral to a national park/reserve where restrictions are placed upon resource use or special development measures are undertaken to enhance the conservation value of the area (Oldfield 1988). A socioeconomic approach is illustrated by the World Bank definition: a social agreement or contract between the protected area and the surrounding community, where size, position, and type of buffer zone are defined by the conditions of this agreement.

Buffer zones aim at controlling human activities within the lands adjacent to a core protected area by promoting their sound management, thus decreasing the potential impacts and the probability of isolation. The presence of a local population is implicitly permitted within the buffer zones (otherwise the buffer zones would be a totally protected area). The current approach in buffer zone design tends to accept them as areas where a plan of land use regulations is applied rather than as clearly defined areas that could have legal protection. Thus, the buffer zone is (or should be) designed (1) to protect local traditional land use; (2) to set aside an area for manipulative research; (3) to segregate land use like agriculture, but also recreation or tourism activities, from the core area in order to avoid adverse effects, and in this way to support direct site management; (4) to manage adverse effects by putting up a barrier for immediate protection; and (5) to locate developments that would have a negative effect on the core area if they were situated elsewhere.

Ecological Corridors

An ecological network is successful if it sustains biological transition and landscape connectivity at all levels where fragmentation, isolation, and barriers to movements and fluxes are defined. A European ecological network as an international network should integrate a set of mutually compatible national networks, being completed themselves by regional and local networks in each country.

Ecological corridors are various landscape structures, other than patches, of size and shape varying from wide to narrow and meandering to straight, which represent links that permeate the landscape, maintaining or reestablishing natural connectivity. Ecological links between patches have always existed also in natural landscapes. Most obvious are migration routes for birds, ant routes, badger routes, and river corridors for fish migration like for the eel and the salmon. Now most of the ecological corridors are primarily the result of human disturbance regimes. Their density and spatial arrangement change according to the type of land use. Their connectivity varies from high to low. Nowadays nature needs different types of ecological corridors that have a complementary role to play in an interconnected habitat island system. As is the case for all land use, ecological corridors require a planning approach.

The term "corridor" has appeared very early in the literature to refer to long-range dispersal (Simpson 1936). The current use of the term ecological corridor has been recommended by Preston (1960), who has attributed it significant properties in spatial population dynamics, allowing the increase of size and enhancing the chance of survival of smaller populations between preserved patches. The currently expanding field of metapopulation dynamics presupposes that some degree of connectivity exists between the spatially arranged subpopulations within the fragmented landscape.

The nature of ecological corridors and their efficiency in interconnecting remnants and in permeating the landscape depend on the area they originate from and the land use mosaic within which they are embedded and of which they consist: remnants (hedgerows), spot disturbance (railroad and power line strips), environmental resource (streams), planted (shelter beds), and regenerated (regrowth of a disturbed strip) (Forman 1983).

Four types of corridors can be defined (Forman and Godron 1986):

1. *line corridors* — narrow strips of edge habitat, such as paths, hedgerows, and roadsides;

2. *strip corridors* — with a width sufficient for the ready movement of species characteristic of path interiors (for example, a wide power line corridor permitting movement of open-country species through a forest);

3. *stream corridors* — may function as one of the previous two, but which additionally control stream bank erosion, siltation, and stream nutrient levels; and

4. *networks* — formed by the intersection of corridors, this usually resulting in the presence of loops, as well as subdividing the matrix into many patches.

The more complex a corridor is, the more it can be multifunctional. The higher immigration rate that can help to maintain species number, increase population size, prevent inbreeding, and encourage the retention of genetic variation can be judged as the main advantage of corridors (Simberloff and Cox 1987). They also increase the foraging area for wide-ranging species and provide escape from predators and disturbances. Of course they also can have negative influences like the breaking of isolation and exposing populations to more competitive species, the possibility of spreading of diseases, exotic species, and weeds, and disrupting local adaptations, facilitating spread of fire and abiotic disturbances.

To define the need and the criteria for ecological corridors, it is first necessary to define what constitutes sufficient habitat to maintain the population of a species. Landscape features and the marine environment are clearly significant for species. Corridors encompass the particular landscape features and contribute to the overall character of an area capable of supporting such species at favorable conservation status. The following functions are related to the conservation measures required:

1. Ensuring adequate breeding success for viability of populations

2. Allowing expansion of existing populations in their natural habitats

3. Allowing expansion of populations into areas within their natural range currently not occupied

4. To meet the migration and seasonal movement needs of a species

5. Allowing populations across their natural range to be ecologically integrated

Core area sites must inevitably be linked with the wider countryside to allow species dispersion to smaller sites. On the other hand species must have the possibility to colonize empty sites within the core areas if available.

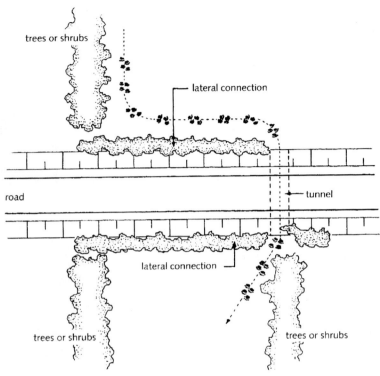

FIGURE 10.2
Design of a road crossing in an agricultural landscape by a tunnel with linking landscape elements. (Ministry of Transport, Public Works and Water Management 1995. With permission.)

Barriers

Barriers can be of all kinds; they are species specific. Increasing traffic and intensifying agriculture made the European cultural landscape more open on the one hand and more closed on the other. A development towards monofunctional land use started in the beginning of this century, and this led to the disappearance of small-scale structures in both agriculture and forestry. Hedgerows disappeared in intensively used agricultural land, forests became uniform production forests, streams have been straightened, and the roadnet became asphalted, more dense, and more intensively used. Last, but not least, many large and important wetlands have been drained. Canalization of waterways and the building of motorways, however, did disturb both the habitat of species as well as their possibility to disperse. That means that in the last century the balance between nature and other land uses has been totally disturbed. Planning of ecological corridors is a method for compensation of a long-term fragmentation process in agricultural landscapes.

The habitat of a species is its abode including the foraging and sleeping sites. These do not necessarily have to be at the same spot. They can even be far apart. In the evening herons, cranes, and storks fly from their foraging sites to their nests that can be kilometers away. The European badger (*Meles meles*) does the same; however, he has to walk and the otter (*Lutra lutra*) has to walk and swim. Bears and lynxes have large habitats in which they move, and an increase in traffic density causes an increase in accidents (Rotar and Adamic 1997). The salmon (*Salmo salar*) has to swim from the sea up a river to reach its spawning grounds in the mountain streams.

Roads are made as technical infrastructures to help human society in its transport needs. Natural infrastructure such as streams and rivers have been adapted to drainage and water transport. Both structure and intensity of use make it impossible for animals to cross these. The structure of roads consists of a wide strip of asphalt or concrete, often with ditches and fences. The structure of waterways consist of straight deep water, weirs and locks, steep shores, and lack of shallow water areas and islands. That makes the man-made infrastructure difficult to cross and for many species it is impossible to reach the other side. Most fishes never get through the maze of locks and weirs in the Dutch delta area.

Planning an ecological networks means also mitigation and compensation of the man-made infrastructure. Fish ladders have to be built to make it possible for fish to cross weirs and locks. Road crossings can be tunnels or fly-overs. Tunnels are used by small species. Habitat elements must be replaced at the right side of the road (Figure 10.2) and they have to be constructed in such a way that wild species are guided towards the tunnel. Fly-overs or ecoducts are meant for larger species (Figure 10.3). In all cases, the landscape

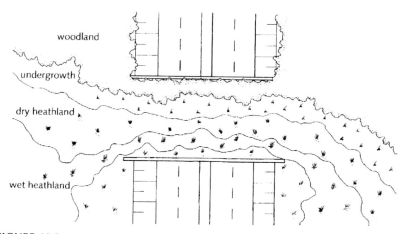

FIGURE 10.3
Ecoduct in a forest and heathland area with several habitat types. (Ministry of Transport, Public Works and Water Management 1995. With permission.)

in its surrounding has to be adapted to its function: hedgerows and small forests for guidance and shelter have to be planted. For those animals using water as a corridor (otter, *L. lutra*) bankside waterway crossings have to be developed. Natural banks must be maintained, and where roads cross waterways tunnels have to consist of both a dry and a wet passage for fauna.

In landscapes where multifunctional land use is required, for instance where outdoor recreation and nature use the same space, a well-designed structure including physical barriers for man can help to construct quiet ecological corridors alongside trails. An ecological corridor for the otter (*L. lutra*) needs shelter and availability of a stream. The trail should be close to nature to allow walkers to enjoy nature, but the shelter of the natural species should not be influenced. In the Dutch lowlands this is done by designing trails and ecological corridors to provide eye contact, but also to prevent preventing physical contact (Figure 10.4).

Implementation of Ecological Networks

Planning for the future is always planning for uncertainty. This is real life in economics, in weather forecasting, but also in planning of ecological networks. It is impossible to know the landscape ecological system, although we might strive towards a better understanding, that makes it possible to give reliable estimates of possible trends in ecological developments in the European landscapes.

Within and outside the European Union ecological networks are being implemented. Every country has its own history to build on, a history of land use, nature conservation and social and political organization. All these aspects influence if and how nature conservation, policy develops into a land use planning strategy. This leads to differences in implementation, differences in legislation, in instruments, and in problems to be solved. Two case studies can give an impression of these differences.

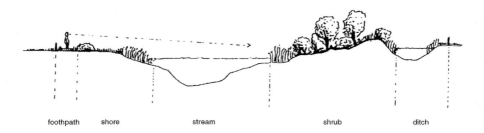

foothpath   shore   stream   shrub   ditch

FIGURE 10.4
Combination of a trail and an ecological corridor in an agricultural landscape. (Elzinga and van Tol 1994. With permission.)

The Netherlands

In the Netherlands the area of nature has been in decline by land development for agriculture during the last 100 years resulting in a potential time lag in species decline. Present land use and infrastructure development enforced this threat by causing isolation of the leftover remnants of nature. Using recent landscape ecological knowledge the concept of a national ecological network was introduced in nature conservation policy. It was decided that the network should be built up in four categories: core areas, nature development areas, ecological corridors, and buffer zones. Core areas had to be of national or international significance. Not all areas included in the Dutch ecological network are of high natural value now. Areas offering realistic prospects for the redevelopment of nature are included in the ecological network as nature development areas.

Ecological networks have been developed at the national level (Ministry of Agriculture, Nature Conservation and Fisheries 1990), but realized at the provincial and local levels. The green network of the Dutch province of Noord Brabant is comparable to the national ecological network, but functions at the regional level. It is not only part of the nature conservation plan, but also a basic report for the regional development plan. It is a spatial coherent structure that covers all more or less natural areas within the province with all kind of functions, forestry nature conservation, and outdoor recreation (trails and attraction areas). Its core is the ecological network consisting of forests, nature reserves, environmentally sensitive areas, nature development areas, and ecological corridors that exist already or will be realized in the near future. The green network is multifunctional; the ecological network within it has only one function.

In The Netherlands, the landscape is characterized by small nature areas and relatively small farms (average 30 ha). Plans have been developed to realize ecological networks. The question, however, is how realistic are the plans? Van den Aarsen (1994) carried out an ex ante-evaluation of the policy plan concerning the green network in the catchment areas of Beerze and Reusel, two lowland streams in Noord Brabant (Figure 10.5). The pivotal question in this research was whether or not the proposals for core areas, corridor zones, and nature development areas are adequate in terms of sustainable conservation of the desired ecosystems. Here intensive agriculture is mixed with small areas of nature conservation interest.

As far as possible, the level of environmental quality that can be realized on the basis of proposed policy measures was compared with the level of quality necessary to ensure the persistence of nature ecosystems. The results of this comparison show a great gap between measures and conditions; some ecological conditions have not been taken into account at all. The regional plans do not lead to sustainable conservation of existing nature ecosystems, and the realization of both adequate corridor zones and nature expansion areas in stream valleys seems doubtful.

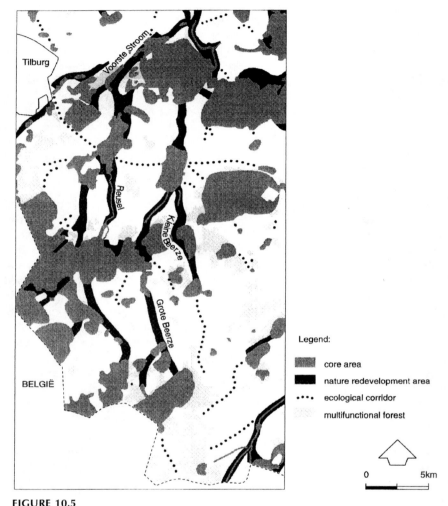

FIGURE 10.5
The Beerze-Reusel area in the province of North Brabant. Indicated are the lowland streams, the proposed elements of the ecological network, and the green network. The white area is agricultural land. (Van den Aarsen 1994. With permission.)

Meeting the conditions for the quality of nature ecosystems would seriously affect the outlook for agriculture in intensive farming areas like the Beerze-Reusel area. Respecting the preconditions for the persistence of nature ecosystems will have far-reaching consequences for the agro-ecosystems. Besides a reduction in acreage, these consequences include the necessity to reduce the release of nitrogenous compounds and phosphate and the use of groundwater, and moreover the creation of stepping stones and corridor zones requires space. This could lead to the necessity of reducing the number of animals in the area.

The extent to which agro-ecosystems in the Beerze-Reusel area should be confronted with restrictions, from the intention of realizing an ecological network, depends on their location with respect to the nature ecosystems concerned. For example, a reduction of the local emission of ammonia is especially necessary in the zones of 500 to 1000 m downwind of susceptible nature ecosystems; restrictions in the application of manure are important in those agricultural areas from which groundwater and surface water flow towards nature areas. On a local scale, a cumulation of restrictions can be expected, because, in order to realize the ecological and green network, all relevant boundary conditions should be taken into account.

Another question is whether or not policy makers have anticipated and accepted these consequences for agricultural land use. In other regional plans, the Beerze-Reusel area has also been indicated as an important agricultural area. To counteract unwanted consequences, the design of a green network leaving adequate perspectives for agriculture requires tailor-made regional and/or local plans. From this point of view, special knowledge of the spatial relationships between agro-ecosystems and nature ecosystems based on their spatial arrangement is imperative to regional planning.

Finally, at least in The Netherlands there are nearly no means to stimulate extensification of agricultural production except of the policy on Environmentally Sensitive Areas and the policy to stimulate biological farming. That means that the future scenario given by the proposed ecological networks give a nice outlook at a better future, but that at least for the intensive farming areas the road to get there is still under construction.

The Czech Republic

In the Central and Eastern European countries integrated planning has a strong tradition. An example is the Czech Republic. Nature conservation as an independent policy had a minor position, but it developed in the framework of physical planning and other forms of integrative planning. In this framework ecological networks consist of several hierarchic levels and is named the Territorial System of Landscape Ecological Stability (TSLES). Most of the networks are part of a systematic approach. Only a few of them on the local level are in a process of physical realization in a landscape. It is especially a problem of a lack of money for realization and also an organizational problem in connection with large changes in ownership. Realization and implementation of the territorial system of landscape ecological stability have their basis in new Czech environmental and nature conservation law of 1992 that mentions the ecological network as the core of Czech nature conservation policy.

The concept of territorial systems supporting landscape ecological stability (TSLES) was already developed in Czechoslovakia in the 1980s (Miklós 1996; Buçek and Lacina 1992). TSLES is built by a network of ecologically important landscape segments purposefully located according to functional and spatial criteria. The present set of relatively ecologically stable segments—disregarding their functional relations—is considered to be a skeleton of landscape ecological stability. TSLES is therefore formed by selecting the most valuable parts of the present landscape with adding missing segments. The area of local biocenters is proposed to be at least 1 to 5 ha, depending on the type of habitat (the minimum area for water habitats is 1 ha, minimum area for forest and grass communities is 3 ha); the minimum width of the local biocorridor is 10 to 20 m, according to the types of communities.

Differences occur in the possibilities for developing an ecological network depending on other land use activities. In the Czech Republic there is a rather clear separation between the intensive farming zone and the areas of low intensive farming. There is now the beginning of the national policy on rural areas dealing with the recent problems.

The land use policy of the Czech Ministry of Agriculture includes subsidizing afforestation on agricultural lands and conversion into grassland. But great regional differences in land use development, depending on natural conditions, especially on soil fertility and land suitability for agricultural production, will exist. Only in the most productive agricultural regions like the Bohemian and Moravian lowlands, which are able to compete within Europe in food production, can intensification of agricultural production be expected. Changes in ownership directed to privatization of the land have not yet changed the main features of the Czech rural landscape till now. Large collective open fields and low landscape stability prevail at least in the major farming areas.

The official environmental policy of the Ministry of Environment and the Ministry of Agriculture of the Czech Republic are now beginning to be directed toward the reduction of subsidies on agricultural production and slowly to support landscape multifunctionality and to keep traditional settlements in the rural landscape. Based on the set-aside programs and with a decrease of agricultural production there is hope for a strengthening and reconstruction of the landscape stability and its natural and environmental values. Serious land use and landscape structure changes are expected in the near future. It is estimated that in the Czech Republic about 15% of the total agricultural land will be available for set-aside programs.

Some segments of TSLES might only be selected on local levels (about 2 to 3% of the total land), especially the intensive farming area, and the farmers shall have to be financially compensated for it. On less fertile soils and in less favorable climatic conditions it will be easier to develop an ecological network. Especially on zones of drinking water production, protected areas for water management, which are concentrated in highlands and submountainous regions, the area of seminatural grasslands can be extended and reach about 30% of the total land, and in this way it will have positive effects on water quality, landscape stability, and erosion. In the mountains above 600 to 700 m above sea level marginalization of agricultural land is resulting in disappearing of characteristic features and the aesthetic values of the traditional rural landscape. One of the solutions is the use of state subsidies in these regions for maintaining landscape, keeping aesthetic and environmental values, and supporting human activities like agrotourism.

Toward an Ecological Network for Europe

Landscape ecological research, the development of nature conservation, and European cooperation as expressed in the Habitats Directive and the Pan-European Strategy brings us to think of a coherent European ecological network. The change in nature conservation strategy in the last decades of the 20th century took place in both in east and west. Planning of ecological networks within a wide diversity of planning systems and politically different systems are the result. The Czech and Slovak systems of landscape ecological stability, the Dutch Ecological Main Structure, the Estonian System of Compensative Areas, the German Ökotopverbundsysteme, and the Danish Naturverbindseler all use principles from landscape ecology. They consider the landscape as a dynamic system that is in continuous change: fast changes in river systems and slow changes in mountains and mires. There is a perspective to merge the national and regional networks into one system without denying the political differences. Scenarios on nature conservation can take into account all the differences that are found within Europe: ecological, social, economic, and political.

One of the policy objectives in Europe for the future must be to conserve nature, landscapes, and open spaces within a reestablishing Europe where new economic developments are immanent and where boundaries become less and less important. We must be able to develop Europe in such a way that there is potential for urban development, transport system development, agricultural development, and forestry and where there is still enough space left for nature. It appeared possible to develop a scenario for nature conservation for the European Community claiming about 30% of the European territory (Bischoff and Jongman 1993). Confrontation with claims for agriculture and forestry showed that this seemingly enormous claim can become reality without great social and economic conflicts (Netherlands Scientific Council for Government Policy 1992).

Within a claim for nature for Europe there are ecological differences. There are biogeographical differences; it is not correct to compare the Irish flora with the Portuguese flora, because one is Atlantic and the other Mediterranean. Due to its history, climate, and geomorphology the Portuguese flora is richer. The importance of the Irish flora is in its heathlands and bogs. However, there are gradients in vegetation as, for instance, from the Irish heathlands onto the Portuguese (Atlantic) heathlands and Mediterranean shrubs. These gradients show that nature in Europe must be approached as one system with many gradients and diversity on all hierarchic levels.

There are social and economic differences within Europe. The border between the Slovak Republic and Poland can easily be observed from the air: the parcel size differs about 100 times. In the Slovak Republic there was a land reform in the 1960s and in the south of Poland there was no land reform. That causes differences in social and economic development and perspectives. We do not have the right to say to the Polish farmers that they have such a rich nature and that they should not improve their land. However, we cannot deny that species richness is concentrated in these landscapes and that in Europe conservation objectives can best be realized there. But also here social and economic development goes on. That is why we have to think of ways of valuing and compensating. Otherwise we did not learn from the past in Western Europe.

Heathlands in The Netherlands are considered to be nature, and heathland management is considered as nature management now. The Dutch pay for heathland management and it is expensive. In northern Portugal heathland is the common grazing ground (baldios) that is used to graze the flocks of the village. Changing this into a forest or a nature reserve is a disruption of the social and economic structure of the region. It also means disappearance of natural species and characteristic habitats. Should we wait until the Portuguese and the Polish are rich enough to pay for the expensive nature management? Or should we be prepared to think of better ways to conserve this agricultural land and nature now. Planning of nature will be as important as the planning of agriculture, infrastructure, and urban areas. That means that a European scenario must be built with the help of a diversity of planning methods that vary between countries and within countries. It requires not only a physical network, but also a planning network and a network of cooperating politicians, planners, land users, and scientists. It also requires knowledge on sustainable land use and valuating economic and ecological gains and losses at all levels. The European ecological network is built from these local, regional, and national bricks.

11

A Land Transformation Model for the Saginaw Bay Watershed

Bryan C. Pijanowski, Stuart H. Gage, David T. Long, and William E. Cooper

CONTENTS

Introduction

A suite of complex factors, including policy, population change, culture, economics, and environmental characteristics, drive land use change. Land use change is one of the most critical dynamic elements of ecosystems (e.g., Baker 1989; Richards 1992; Riebsame et al. 1994; Bockstael et al. 1995). Human-induced changes to the land often result in changes to patterns and processes in ecosystems such as alterations to the hydrogeochemistry (Flintrop et al. 1996), vegetation cover (e.g., Ojima et al. 1994), species diversity (Costanza et al. 1993), and changes to the economies of a community. It is for these reasons that issues surrounding land use are central to the concerns of local and regional resource managers and community land use planners.

Information about current land use patterns, the causes of land use change, and the subsequent effects of these changes can be effectively communicated to resource managers, community planners, and policy analysts using geographic information systems, predictive models, and decision support systems (Cheng et al. 1996; Doe et al. 1996). The advancements in many geographic information system applications such as ARC/INFO (Environmental Systems Research Institute 1996) and the increased accessibility of spatial databases makes developing simulation models within geographic information systems more feasible than even a few years ago.

This paper presents an overview of the modeling framework, systems approach, and spatial class hierarchies of our pilot, GIS-based Land Transformation Model (LTM). Our LTM has been developed to integrate a variety of land use change driving variables, such as population growth, agricultural sustainability, transportation, and farmland preservation policies for the Saginaw Bay Watershed (SBW) in Michigan. The pilot LTM utilizes a set of spatial interaction rules, which are organized into an object class hierarchy. The model is entirely coded within a geographic information system with graphical user interfaces that allow users to change model parameters. Output of the LTM includes a time series of projected land uses in the watershed at user-specified timesteps.

Project Objectives

The objectives of the Land Transformation Project are to: develop a spatial–temporal model that characterizes land use change in large regions; create a model that is transferable in scope to other regions undergoing land transformation; incorporate policy, socioeconomics, and environmental factors driving land use change; develop a pilot LTM that demonstrates proof of concept and that can be used to generate spatial and temporal aspects that can be generalized for the development of new model components; apply a systems approach to model development; and use the model to test "what-if" policy scenarios.

Conceptual Elements

The LTM (Pijanowski et al. 1995, 1996, in review) describes the influence of land use change on ecosystem integrity and economic sustainability of large regions. Conceptually, the LTM contains six interacting modules (Figure 11.1): (1) Policy Framework; (2) Driving Variables; (3) Land Transformation; (4) Intensity of Use; (5) Processes and Distributions; and (6) Assessment Endpoints. All modules and submodules within the conceptual diagram are recognized not to be mutually exclusive; we use this diagram to illustrate main points and provide a foundation for the description of more detailed model components. The pilot LTM that is described below contains two of the six LTM modules, driving variables and land transformation. The spatial extent of the LTM can be any definable region; however, because future model developments will be focused on coupling land use change and hydrogeologic and geochemical processes, we give precedence to watersheds as the spatial extent in LTM applications.

The Policy Framework module of the LTM organizes the goals for the stakeholders of the watershed who include resource managers, private and corporate landowners, and local land use planners. Stakeholder goals may include: control of pollutant inputs, ecological restoration, habitat preservation, improving biodiversity and biological integrity, and facilitating economic growth. Within this framework, many stakeholder goals are under certain types of constraints (e.g., economic, environmental), are made with certain expectations of outcomes, and with specific spatial and temporal scales in mind. For example, a township land use planner is likely to be making decisions within his/her own township. Likewise, a state or federal government resource manager might be concerned about areas that encompass several counties.

The LTM contains three general categories (Figure 11.1) of Driving Variables: Management Authority, Socioeconomics, and Environmental. Management Authority includes the institutional components and policies of land use. Land ownership is an important component in this module of the model since state and federally owned lands (e.g., state and federal forests, parks, and preserves) need to be excluded from development. Socioeconomic driving variables include population change, economics, of land ownership, transportation, agricultural economics, and locations of employment. Environmental driving variables of land transformation are (1) abiotic, such as the distribution of soil types and elevation, and (2) biotic, such as the locations of endangered and threatened species, or the attractiveness of certain types of vegetation patterns in the landscape for development. Driving variables may

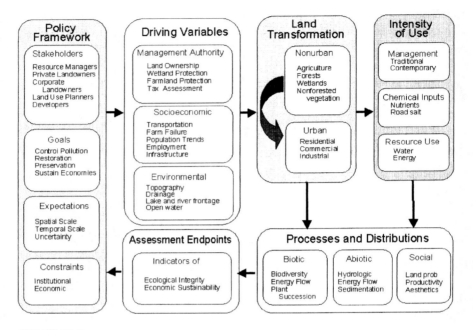

FIGURE 11.1
Conceptual elements of the land transformation model.

contain intercorrelated subcomponents; hence the model can be hierarchical. For example, the farming socioeconomic system in the SBW application of the pilot model is composed of farm-size dependent economics, farmer demographics, and environmental influences on farm productivity.

Land Transformation is characterized by change in land use and land cover. Land use describes the anthropogenic uses of land as it affects ecological processes and land value (Veldkamp and Fresco 1996). Land uses that we consider at the most general level are urban, agriculture/pasture, forest, wetlands, open water, barren, and nonforested vegetation. Land cover characterizes the plant cover of associated land use and is thus not mutually exclusive of land use. Land cover types that are considered include: types of agriculture (row crops vs. nonrow crops), deciduous and coniferous forests, and nonforested vegetation.

Within each land use, we consider Intensity of Use such as land management practices, resource use, and human activities. Intensity of use can be measured as chemical inputs to the land to increase its productivity (e.g., herbicides), chemical inputs as it results from human activities (e.g., salting of roads), and natural resource use (e.g, subsurface water for irrigation, per unit area energy consumption, and forest harvesting). Socioeconomics, policy, and environmental factors will also drive the intensity of use as well.

Changes in land use and cover and intensity of use alter Processes (e.g., hydrogeologic and geochemical) and Distributions of plants and animals in ecosystems. Processes that we are interested in characterizing include groundwater and surface water flows, chemical and sediment transport across land and through rivers and streams, geochemical interactions, and fluxes such as nutrients (nitrogen and phosphorus). Land use and land cover will affect the types and numbers of animals inhabiting areas.

Assessment endpoints are indicators of ecological integrity and economic sustainability. These assessment endpoints are used to quantify the nature of changes in landscapes. It is important that assessment endpoints be (1) relatively easy to quantify, (2) unambiguous, (3) correlated with changes to land use, and (4) reflect qualitative aspects of landscapes. These assessment endpoints provide input to the decision-making process by watershed stakeholders.

Spatial Framework

Land use and features (roads, rivers, etc.) in the watershed are characterized in the pilot LTM model as a grid of cells. Each cell is assigned an integer value based on land use (e.g., urban, agriculture, wetlands, forest) or land feature. Driving variable calculations produce land use conversion probabilities for each cell. The geographic information systems (GIS) is used to perform these driving variable calculations, integrate all driving variable conversion probabilities, and produce future land use maps for the entire watershed. GIS calculations in grids commence at the upper left corner of the grid and end at the lower right corner of the grid. In the SBW application of the pilot LTM, up to 5.2×10^7 cells, are contained in each grid.

Figure 11.2 illustrates conceptually how land use transitions are determined in the LTM. This hypothetical landscape contains three agricultural parcels: a small parcel near a highway, a large parcel some distance away

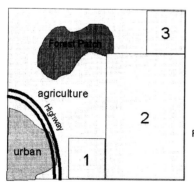

Farm #3

far from urban infrastructure	relative prob = 3	weight = 4
not profitable due to size	relative prob =10	weight = 3
old farmer ready to retire	relative prob = 9	weight = 4
tax assessments rel . low	relative prob = 3	weight = 2

Moderately high transition prob (prob = 84; percentile class = 91)

Farm #2

far from urban infrastructure	relative prob = 6	weight = 4
profitable due to size	relative prob =1	weight = 3
corporate farm	relative prob = 1	weight = 4
tax assessments moderate	relative prob = 6	weight = 2

Low transition prob (prob = 4 3, percentile class = 23)

Farm #1

near urban infrastructure	relative prob = 10	weight = 4
not profitable due to size	relative prob =9	weight = 3
aging farmer	relative prob = 1	weight = 4
tax assessments high	relative prob = 10	weight = 2

Very high transition prob (prob = 91, percentile class = 99)

FIGURE 11.2
Relative land transition probabilities.

from the highway, and another small parcel a relatively large distance away from the highway. The drivers to land use change operate on these parcels differently depending upon the spatial relationships of the parcel and the drivers. For example, parcel #1 is under pressure for development due to its proximity to a highway, proximity to urban infrastructure such as city water and sewers, proximity to high-density employment centers found in the urban areas, and, due to its size, the farm is not likely to be profitable. Furthermore, its landowner may also be older and because few younger people are entering agriculture, it is at a high risk of being converted out of agriculture and into an urban use. The second farm, as indicated by parcel #2, is held in agriculture by the nature of its ownership (i.e., corporate). Parcel #3 in this figure has a higher probability of converting to urban land use because of the demographics of the owner and the size of the parcel.

In the LTM, we use the GIS to make spatial calculations between drivers of land use change and cells being considered for land transition. The values resulting from these calculations are converted to relative land transition probabilities. Relative land transition probabilities that are used range from 1 (lowest probability of undergoing transformation to urban land use) to 10 (greatest chance of being converted to urban land use). Creating these relative probabilities from absolute GIS calculations requires (1) spatial scaling or assigning relative transition probabilities based on absolute values and (2) making adjustments to state transition patterns. The types of spatial scaling and state transitions considered in the LTM are described below as part of the presentation of spatial class hierarchies.

In addition to calculating relative land transition values based on (1) spatial interactions of drivers and (2) cells within a parcel, relative weights for each driving variable are assigned, and these are then used to calculate urban transition values for each cell in an area. All land transition probabilities and weights for each driving variable are then integrated with the GIS for each location. Values are then placed into equal area percentile classes. Cells with the greatest percentile value are assumed to transition first to urban. The number of cells for each future transition is based on the per unit area requirements for urban given population growth projections for an area (township, county, or entire region). The number of cells that meet the demands for each successive projection (e.g., decades) are then transitioned to urban. A more detailed description of the model calculation process can be found in Pijanowski et al. (in review).

Spatial Class Hierarchies

Figure 11.3 illustrates the LTM Spatial Object Class Hierarchy. There are six principal spatial classes in the LTM: interactions, resolution, spatial scaling, state transitions, landscape features, and the number of subdrivers. Each of the principal spatial classes in turn are composed of several subclasses, which may be further divided into more refined spatial objects. The terminal posi-

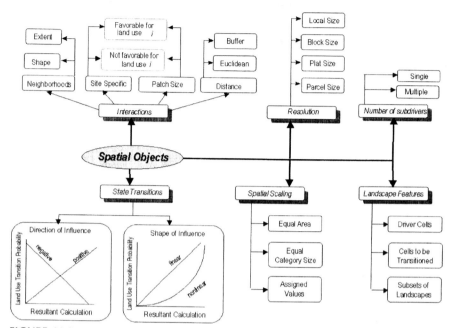

FIGURE 11.3
Spatial objective class hierarchy.

tions of the space object classes become rules from which software modules are developed within the geographic information system.

Spatial Interactions

Spatial interactions used in the LTM are neighborhood, distance, patch size, and site-specific characteristics. Neighborhood spatial interactions are based on the premise that trends and patterns in neighboring locations influence the land use transition probability of a cell. Neighborhood interactions can also vary in size, from those that only occur among proximal locations to large neighborhoods that encompass large areas (counties, subwatersheds or the entire watershed). We also recognize that the shape of a neighborhood may differ, from square or circular to irregular (e.g., watershed catchment). Distance functions are the second type of spatial interactions used to characterize driving variables of land use change. We use the GIS to calculate the distance of locations in the watershed from landscape features (e.g., roads, rivers, employment centers) and convert these "raw" values into relative probabilities of land transformation (conversion rules are described under state transitions below).

Patch Size is based on the principle that the size of a parcel of land held by an owner has an influence on whether a land use conversion is imminent. For example, farm size in the U.S. impacts profitability such that small farms cannot compete with larger farms that can invest in advanced farm machinery, etc. Thus, small farms are at greater risk of failure and hence being converted to a nonagricultural use than larger farms.

Site-specific characteristics are also important to land use conversion. Certain characteristics (e.g., soil type or elevation) make each site suitable or unsuitable for a particular land use. Policy may also influence site-specific characteristics of land transformation by either "locking" land in a specific land use or "promoting" its conversion.

Resolution

Examples of the resolution spatial object class in the LTM include those for cell size. Four different resolution classes are used in the LTM: parcel (30 × 30 m), plat (100 × 100 m), block (300 × 300 m), and local (1 × 1 km). These rules were developed to characterize certain processes such as land ownership changes which occur at relatively high resolutions (e.g., 30 × 30 m) and hydrologic dynamics that occur at more coarse resolutions (e.g., 1- × 1-km resolution). Selection of resolution is also determined by the resolutions of databases available to study a process or pattern (e.g., land use is 30 × 30 m as it might be developed from Landsat™). We integrate multiple grids using our GIS.

Spatial Scaling

Creating these relative probabilities from the absolute GIS calculations requires (1) spatial scaling or assigning relative transition probabilities based on absolute values and (2) making adjustments to state transition patterns. Spatial scaling to convert all "raw" GIS calculations (e.g., distances) to relative probabilities is accomplished using the slice function in ARC/INFO GRID. Two options of this function are employed: equal area or equal class sizes. The former option of the slice function produces driving variable grids with equal numbers of cells with values between 1 and 10. The latter option provides driving variable grids with equal size classes between the largest and smallest values in the entire grid.

Relative transition probabilities can be assigned based on absolute values rather than using spatial scaling routines as described in the previous paragraph. For example, in the SBW application of the pilot LTM, relative transition probability values of 10 were assigned to all cells 30 m on either side of state and county roads within 100 m of highway intersections; all cells 30 m around county and state intersections were assigned values of 7; and all cells on either side of state and county roads were assigned values of 5.

State Transitions

Two different state transition adjustments were made in the LTM. First, the direction of the relationships between the spatial scaling routine result and land transition probability may be positive or negative. For example, land closer to road intersections has the greatest probability of conversion to urban. The GIS is used to calculate the Euclidean distance of cells from the nearest road intersection, and these values are then spatially scaled to create grids with relative probability values where the largest values are assigned 10 and the smallest values a 1. However, land closest to a driver such as a road has the greatest probability of conversion to urban; thus, there is a negative relationship between the result of the spatial scaling and the degree of urbanization. We "invert" these transition values using the following simple expression:

$$outgrid = 11 - ingrid$$

where outgrid is the inverted driving variable grid and the ingrid is the input grid that contains values from 1 to 10.

The relationship between a spatial calculation and the influence of this result on urbanization can also be linear or nonlinear (Figure 11.3). The equal size class option of a slice is only used for spatial scaling of these state transitions.

Landscape Features

The fifth type of spatial class objects in the LTM are landscape features. In many instances, the presence or absence of a feature in the landscape is important in the calculation of a land transition probability. For example, the relative density of farms in a local area are derived by producing a map of the presence (coded as 1) or the absence thereof (coded as 0) of agriculture in all locations in the watershed. Features are also cells that are considered for transition and those that are drivers of land use change.

Number of Subdrivers

Single or multiple layers are required to develop a driving variable. Multiple layer examples include those subdrivers that influence farm failure such as farm size, farmer age, amount of available surrounding arable land, soils, climate, and farm infrastructure (e.g., drains).

GIS Framework

GIS Integration Schematic

Figure 11.4 illustrates how the GIS is used to produce land use projection maps. The first step is to create driving variable grids that contain values representing relative transition urban probabilities. This process may first require producing grids that contain information about the absence (cell value = 0) or presence (cell value = 1) of a feature (e.g., road) or land use type (e.g., agriculture); several grids may be integrated to produce the necessary input layer (Figure 11.4, Step 1A). Spatial calculations (e.g., neighborhoods,

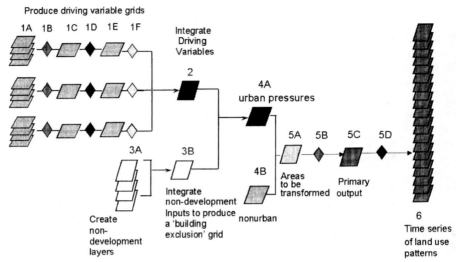

FIGURE 11.4
GIS integration schematic.

Euclidean distances) are performed (Figure 11.4, Step 1B) on the input grids so that resultant "raw" values (e.g., distance a cell is from a driver cell) are stored in each cell in the grid (Figure 11.4, Step 1C). These "raw" values are then scaled (Figure 11.4, Step 1D) so that there are an equal number of values between 10 (greatest probability on urbanization) and 1 (least probability for urbanization). This process produces driving variable grids (1E) that are then multiplied by a driving variable weight (1F). All driving variable grids are then summed (i.e., all cells for each location are added together) and this sum is stored in a final integrated driving variable grid (Figure 11.4, Step 2).

Cells within the grid that are identified as nonbuildable due to policy (e.g., development rights have been restricted) or ownership (e.g., land is state or federally owned) are created (Figure 11.4, Step 3A) so that no-buildable cells are assigned value of 0 and potentially buildable areas assigned values of 1. All of these grids are integrated by multiplying them together so that a single "building exclusion" grid is produced Figure 11.4, Step 3B).

An urban pressure grid is produced as part of Step 4 in the GIS integration process; this is created by multiplying the "building exclusion grid" with the integrated driving variable grid. A nonurban grid (nonurban cell = 1; urban = 0) is used to multiply with the urban pressure grid. This step results in an "area to be transformed grid" (Figure 11.4, Step 5A) that contains integrated driving variable values for all nonurban areas. Values in the nonurban areas are then scaled (Figure 11.4, Step 5B) into percentile classes so that each percentile is represented by an equal number of cells (i.e., each value between 1 and 100 contains equal areas) in the grid labeled as 5C. The number of cells transformed to urban is determined by calculating a "critical threshold value" (Step 5D). Estimating the appropriate critical threshold value can be accomplished as follows. First, the amount of future urban land is determined using population growth projections and per capita urban land requirements:

$$U(t) = \left(\frac{dP}{dt}\right) * A(t) \tag{1}$$

where U is the amount of new urban land required in the time interval t, P is the number of new people in any given area in a given time interval, and A is the per capita requirements for urban land. The critical threshold value is then simply a proportion of the current nonurban land use to the amount of new urban land use required in the future:

$$C(t) = 100 - \left\{[U(t)/N] * 100.0\right\} \tag{2}$$

where N is the amount of current nonurban land use that can be developed in the future, expressed as a percent. Note that N is also a function of non-buildable area. Future land use grids are produced that "step" through the critical threshold values (Step 6).

Model Interface

Figure 11.5 shows a sample user interface of the LTM. This interface was developed using the Formedit GUI development tool in ARC/INFO in the OpenWindows UNIX environment. The interfaces allow users to set values for driving variable calculations (e.g., cell size, neighborhood extent) as well as provide access to visualization and output analysis tools.

Model Application

Site Description

The Saginaw Bay Watershed (SBW) is one of the largest watersheds in the Great Lakes area (Figure 11.6) covering approximately 15,000 km² (15% of the total area of the state of Michigan). The SBW is composed of 10 smaller watersheds which are further divided into 69 subwatersheds. The principal river in the watershed is the Saginaw, which is only 47 km long; however, it drains 28 rivers and streams and nearly 73% of the watershed (MUCC 1993). There are three major tributaries of the Saginaw River: the Cass River to the east, the Flint River to the south, and the Titabawassee River to the west. The major cities within the watershed include Flint, Saginaw, Bay City, Midland, and Mt. Pleasant. There are 22 counties, 42 cities, 50 villages, and 277 townships in the watershed. Each municipality (e.g., township or cities) is given the authority to govern their own land use. Over 1.1 million people live in this watershed.

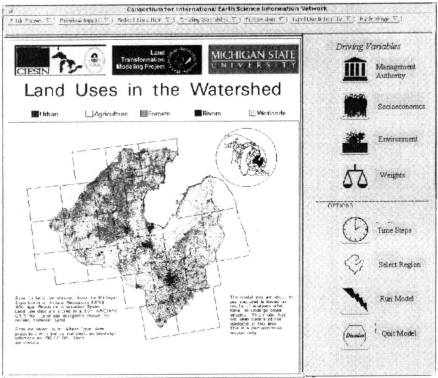

FIGURE 11.5
Sample graphical user interface of the pilot LTM.

Agriculture is by far the most common land use in the SBW (46%), followed by forested areas (27%), and open vegetation (nonforested vegetation) (11%). In the SBW, fewer than 8% of the cropland is under conservation tillage compared to the statewide average of 40% (MUCC 1993). The lack of conservation tillage practices has created a situation of massive soil erosion due to wind and water (MDNR 1994). Urban use makes up only 6.6% of the entire area. Within urban areas, residential areas comprise 67% of the urban area. The other major urban uses are commercial (9%), transportation (8%), and industrial (4%). Topography does not vary considerably in the watershed. Areas near the mouth of the Saginaw Bay differ by less than 3 m from 10 mi inland. As a result, flow of the major streams in the Saginaw is relatively slow; in some cases, the Saginaw River has been known to flow in the reverse direction during strong northeasterly winds.

Pilot LTM Driving Variables

We have used the LTM conceptual diagram (Figure 11.2) to develop a pilot GIS-based simulation model that forecasts land use in the SBW using policy,

FIGURE 11.6
Saginaw Bay Watershed.

socioeconomic and environmental driving variables. This model represents two of the six LTM modules. The driving variables of this pilot model are land ownership, the state's farmland preservation act and its effect on farm to urban conversion, the state's wetland protection act, the effect of the state's property tax assessment method on farm failure, the Suburban Control Act, local and regional population change, economics of land ownership, transportation effects on urbanization, local and farm-level agricultural economics, location and density of employment opportunities and social factors that affect farm failure, the presence or absence of buildable soils, the effects of drainage system on agricultural performance, and the relative attractiveness of several landscape features for urban development. Figure 11.7 illustrates some of the driving variable calculation results. A more detailed description of the driving variable calculation formulation can be found in Pijanowski et al. (in review).

Results and Discussion

The pilot LTM was executed without assigned weights to the 13 driving variables listed above (Figure 11.8). The critical threshold values for each 10-year

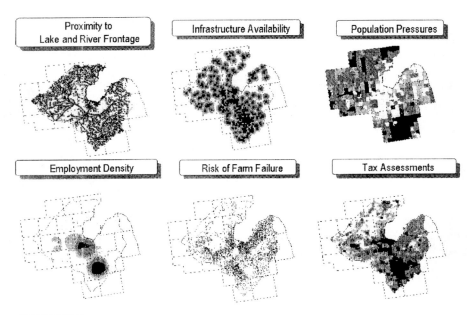

FIGURE 11.7
Sample pilot LTM driving variables.

timestep was determined from State of Michigan population projections for the next 50 years. In addition, the base land use map that was used was developed by synthesizing land use polygons from 350 townships in the watershed from the Michigan Resource Information System. Land use from this database is current only to 1980. Thus, the first projection created a land use map for 1990. In the near future, the pilot LTM will be calibrated by conducting an historical forecast in order to attempt to predict current land use conditions. A current land use map for the entire watershed is planned to be completed by August of 1999.

Currently, we build model animations of outputs to examine how various model parameters affect model outcomes. Typical animations include annual timesteps of model forecasts for the entire watershed and for subareas. This method, first employed by Clarke et al. (1997) for their land use change model, is very powerful because many different types of output patterns emerge from animations.

Many spatial models that have been developed currently do not utilize a GIS for simulation. For example, the Spatial Modeling Environment of Costanza (Costanza et al. 1990, 1993) uses a GIS for visualization of the final outputs of their spatially explicit landscape model. The spatial modeling is accomplished using an object-oriented framework, the STELLA process-based modeling environment, and the C++ programming language. There are many advantages to using GIS to model spatial dynamics, however. First, many of the data layers for spatial models already reside in a GIS and hence are easier to manage if they stay in a GIS (Maidment 1991, Ball 1994). Second, many GIS packages already contain the spatial functions required for spatial modeling. In some instances, these functions are very flexible and hence they extend the power of some models. For instance, neighborhood calculations in the LTM can be accomplished using various parameter switches, including setting the size of the extent (i.e., window), the shape of the neighborhood, and the type of neighborhood calculation (e.g. sum, variance, etc.).   A third advantage of modeling within a GIS is that model visualization and analysis are relatively easily accomplished using a GIS. For the LTM, we regularly use spatial data layers that are not part of the modeling input layers to visualize the model outputs. We routinely use spatial overlays, zoom and pan over large areas, and calculate subsets of certain data layers to highlight important areas on which to focus attention. Finally, building the necessary routines using other programming languages can add substantial time to the model development process. New routines will have to be analyzed for programming errors, and this process can take considerable amounts of time.

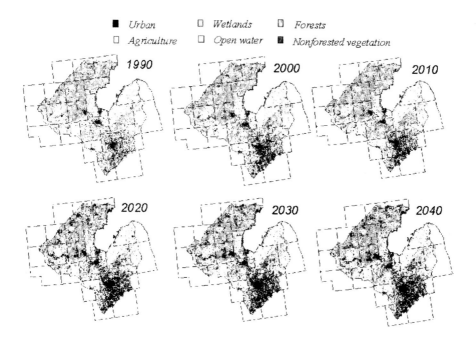

FIGURE 11.8
Sample LTM output for the Saginaw Bay Watershed: land use patterns for the next 50 years.

Acknowledgments

We gratefully acknowledge funding from the Consortium for International Earth Science Information Network (CIESIN), University Center, Michigan, through Cooperative Agreement (CX821505) between the U.S. Environmental Protection Agency and CIESIN. We would also like to thank Bob Worrest, Mike Thomas, and Bob Bourdeau of CIESIN; Charles Bauer and James Bredin of the Michigan Department of Natural Resources for helpful comments during the model development phase. Dennis Tenwolde reviewed earlier versions of the manuscript. Amos Ziegler, Katie Jones, Tom Sampson, Gary Icopini, John Abbott, and Mark Rousseau helped to prepare a variety of the spatial databases used in this project. An earlier version of this manuscript was presented at the National Center for Geographic Information and Analysis (NCGIA) Land Use Modeling Workshop, held at the EROS Data Center, Sioux Falls, SD, June 4–5, 1997.

12

Individual-Based Models on the Landscape: Applications to the Everglades

Donald L. DeAngelis, Louis J. Gross, Wilfried F. Wolff, D. Martin Fleming, M. Philip Nott, and E. Jane Comiskey

CONTENTS

Introduction

Theoretical ecology has long been associated with the use of relatively simple mathematical models to describe populations and communities. These models are descendants of the logistic and Lotka–Volterra models, in that they are differential equations (or difference or partial differential equations) and usually contain some sort of nonlinearity, which acts ultimately to limit populations. There have been many elaborations of these models, such as the inclusion of internal age or size structure (e.g., Metz and Diekmann 1988; Caswell 1989) and the inclusion of spatial extent (e.g., Okubo 1980). However, the basic nature of the models remains the same; mathematical models that are simple enough to be written in the compact form of differential or partial differential equations and analyzed. Such models are referred to in general as state variable models. A state variable is used to represent numbers or densities of organisms of a particular population being modeled, or, alternatively, subpopulations such as particular age or size classes within the population, or subpopulations in particular spatial areas.

During the last three decades, a new modeling approach has developed, individual-based modeling, that is fundamentally different. No state variables are used for population size. Instead, the population is represented as a collection of individuals that are individually modeled (see Huston et al. 1988 and DeAngelis and Gross 1992 for reviews). The focus of the model is on the growth, foraging, survival, reproduction, and other activities of each individual. If one wants to know the total population size, it is necessary only to add up all of the individuals at a given time.

What distinguishes this individual-based modeling approach from the classical models, then, is a different choice of state variables. Individual-based models (IBMs) use variables attached to individuals, individual state variables (ISV), rather than population-level variables to describe the system. The characteristics of each organism (age, size, spatial location, sex, health, social status, experience, knowledge, etc.) constitute the set of variables of the system. Both the number of living individuals and the values of each of their variables can change through time. Such models have long been used to describe a variety of ecological situations. In particular, some of the early work includes models of:

> Interactions between plants and other sessile organisms (e.g., Botkin et al. 1972; Maguire and Porter 1977; Ford and Diggle 1981);
>
> Movement of animals (e.g., Rohlf and Davenport 1969, Siniff and Jessen 1969; Skellam 1973; Yano 1978; Kitching 1971);
>
> Transmission of diseases across populations (e.g., Bailey 1967; David et al. 1962);
>
> Population genetics of small populations (e.g., MacCluer 1967; Schull and Levin 1984); and
>
> Animal behaviors; e.g., flocking behaviors in birds, schooling of fish, spacing in response to spatial distribution of food (e.g., Thompson et al. 1974).

Much of the popularity of IBMs results from their reflection of some basic features of real populations: in particular, each individual within a population is unique and differs from others in many biologically important respects. Such differences are easily accommodated in IBMs. For instance:

> Individuals are capable of complex behaviors, better described by sets of rules for individuals than by equations at the population level.
>
> Populations of higher trophic-level organisms are often small and hence dominated by stochastic variations. These are not easy to incorporate into population-level equations.

Interactions between organisms are usually highly local spatially, which is difficult to represent by simple equations.

Movement of organisms in complex landscapes is more easily and properly described by sets of rules attached to an individual than by equations (e.g., partial differential equations) at the population level.

The majority of early individual-based modeling involved the modeling of plants, either as single species or mixed stands (e.g., JABOWA, FORET, SORTIE). One of the areas of animal ecology where IBMs have been used extensively is in the simulation of young-of-the-year fish cohorts, where the sizes of individuals in the cohort can differ greatly and strongly influence the recruitment to yearlings (e.g., DeAngelis et al. 1992). Another area is that of the interaction of herbivores with patchy spatial distributions of their plant forage (e.g., Cain 1985).

Currently, IBMs are being combined with GIS maps used to describe species populations, including endangered or rare species, on complex landscapes (e.g., Comiskey et al. 1995; DeAngelis et al. 1998). The approach is currently being applied to model several species of the Everglades under a U.S. Geological Survey Program, Across-Trophic-Level System Simulation (ATLSS). This type of approach will form much of the discussion of this chapter.

An article by Levin et al. (1997) recently outlined some of the potential symbiotic interactions of new computational approaches and mathematical analysis in ecology and other areas of ecosystems science. In doing so, however, the authors made statements that should be more carefully considered. Although the review is a useful one, we feel that it misunderstood the way IBMs are being used. In particular, their comments include:

> Because models of this sort may provide an unjustified sense of verisimilitude, it is important to recognize them for what they are; imitations of reality that represent at best individual realizations of complex processes...

> The amount of detail in such models cannot be supported in terms of what we can measure and parameterize... The result is that these models produce cartoons that may look like nature but represent no real systems.

Other papers, such as Wennergren et al. (1995), who assessed the use of spatial models in conservation biology including population IBMs, have echoed the view that available data seldom exist to support development of IBMs.

The discussion in the present chapter will be aimed at describing the application of IBMs to species conservation questions, and to some degree, at answering criticisms that individual-based approaches have engendered by presenting examples of IBMs that are currently being used in modeling animal populations in the Everglades.

Individual-Based Modeling in Applied Ecology

Levin et al. (1997) claim that IBMs "may provide an unjustified sense of veri-similitude." This is a somewhat ironical statement in view of the history of mathematical ecology. If the results of simple models such as the Lotka–Volterra predator–prey model had not had an uncanny resemblance to cycles of fish or lynxes and hares, probably little attention would have been paid them. The Lotka-Volterra type model, which has spawned many variations (e.g., Rosenzweig-MacArthur model) was borrowed by theorists from the equations for chemical kinetics. They were used, less because careful observations of the many animal and plant species suggested them, than because they were mathematically tractable and produced interesting behaviors, including cycles that resembled some well-known cycles in nature.

Thus, resemblance between models and observations has always been a main, if sometimes unspoken, argument for the use of those models in ecology. Seldom, if ever, are the analytic models of population ecology derived in the rigorous fashion that a first-year student of physical chemistry or the physics of fluids must reproduce the derivation of chemical kinetics equations, the diffusion equation, or other equations of those fields. Today in mathematical ecology, the same tradition of justifying models based on resemblance (sometimes superficial) of observations continues. For example, many theorists make much of the fact that models of deterministic chaos can produce output that resembles certain time series population data.

IBMs represent a different approach from the classical models of mathematical ecology. The IBM modeler starts from what is known about the actions of individuals under various circumstances. These actions, even if they are very complex, can be represented through computer simulation. The IBMs start with these mechanisms at the level of individuals and attempt to predict the dynamics that should occur at the population level under given circumstances. An IBM can be applied to populations of arbitrarily small size and in highly non-uniform landscapes.

The verisimilitude that IBMs display is not an accidental factor. A basic feature of the approach is that the models predict patterns at a variety of scales of aggregation, from the individuals up to the population level. This is a conceptual advantage, because the models incorporate causal chains leading from the actions of individuals to total population behavior.

In addition, IBMs are amenable to several levels of verification. One type of validation of models is "face validity," where experts in the subject are asked to compare the patterns predicted by the model with their understanding of the system (Rykiel 1996). This type of validation can be applied to IBMs, because they produce output on the detailed distribution of ISV, including distributions in space. This type of validation also tends to be highly Popperian, as these experts invariably try to find fault with the models in comparing them with what they know of organisms in the field. If the verisimilitude of the models is truly "unjustified," then such a process of validation will detect this.

All models, including very complex IBMs, are certainly abstractions and their usefulness is that they can represent aspects of reality with enough accuracy to help answer questions. But if the verisimilitude that the IBM display extends to making useful predictions, then it is certainly justified.

Levin et al. (1997) further state concerning IBM that "the amount of detail in such models cannot be supported in terms of what we can measure and parameterize...." Wennergren et al. (1995) make the same argument that IBM cannot be supported by data, and that their results are then likely to be erroneous. Wennergren et al. (1995) leave the impression that their negative assessment applies in general to spatially explicit individual behavior models, although their analysis is restricted to a particular dispersal model — a model later published in Ruckelshaus et al. (1997) which was subsequently shown to be in error by two orders of magnitude (see Mooij and DeAneglis 1999).

We are very concerned that notions from papers such as that of Wennergren et al. (1995) and Ruckelshaus et al. (1997), although factually incorrect, are being repeated in the literature. In particular, the view that IBMs are data-hungry and make demands on data accuracy that are impossible to fulfill seems to be widespread. We find this conclusion is of little or no relevance to many of the applications of IBMs to conservation problems. In fact, as will be shown below, IBMs used in applications can be tailored to use spatially explicit empirical data and physiological, behavioral, and natural history information that are typically available from many population and ecosystem studies. Many IBMs are "tactical" models with limited predictive objectives. Data needs for these models are usually parsimonious and can be met with existing or routinely collected data. Other IBMs are more strategic and contain dispersal phases, but without the same degree of sensitivity of model results. Below we consider four examples drawn from our Everglades research.

There are two major components of IBMs as we have used them. The first is a dynamic, spatially explicit description of the landscape. This landscape description includes at least changing water levels at a biologically relevant scale of resolution, 500 × 500 m in this case. Depending on the species modeled, it may also contain vegetation type on the same or finer scale, and a model for changing prey availability.

The second major component is the individual-based description of the species. The models may simulate on this dynamic landscape the entire life cycles of all of the individuals in the population are modeled over many years. Alternatively, the model may simulate the population, or subpopulation, only during the reproductive season. Some models simulate the detailed bioenergetics of individuals, while others may simply simulate demographics and important behaviors, such as movement. This depends on the type of questions being asked and the data available.

Example 1—Cape Sable Seaside Sparrow

The Cape Sable seaside sparrow (*Ammodramus maritima mirabilis*) is an eco-logically isolated subspecies of the seaside sparrow (Beecher 1955, Funder-burg and Quay 1983; Post and Greenlaw 1994). Its range is restricted to the extreme southern portion of the Florida peninsula almost entirely within the boundaries of the Everglades National Park and Big Cypress National Pre-serve (Werner 1975, Bass and Kushlan 1982). The sparrow breeds in marl prairies on either side of Shark River Slough. Marl prairies are typified by dense mixed stands of gramminoid species usually below 1 m in height, nat-urally inundated by fresh water for 2 to 4 months annually. The potential of such habitat for sparrow breeding is dependent upon regimes of fire, hydrol-ogy, and catastrophic events (hurricane and frost).

Recent declines in the sparrow population across its entire range, especially the western portion, highlight the need for an effective ecological manage-ment strategy. The remaining core of the population occupies approximately 60 to 70 km² in the area adjacent to the southeast of Mahogany Hammock. This subpopulation currently represents 73% of the total population (1996 estimate), and because of the spatial restriction it is seriously at risk to the effects of hurricane or wildfire. Changes to the hydrology of the southern Everglades may also increase the threat of extinction. Increased hydroperiods affect the sparrow in two ways: (a) they can directly shorten the potential breeding season and (b) they can affect them indirectly by causing changes in the vegetation. Recent studies (Nott et al. 1998) show that wetter conditions cause typically short-hydroperiod vegetation (*Muhlenbergia*) to become dom-inated by sawgrass (*Cladium jamaicense*) and spikerush (*Eleocharis spp.*). This kind of habitat is less suitable for breeding purposes, but remains available for foraging.

The main objective of the model (SIMSPAR) is to investigate the effects of fire and hydrology regimes upon various measurements of the sparrow pop-ulation. These include lifetime reproductive success of individuals, move-ment patterns and spatial distributions of the population, and fluctuations in the size and structure of the population and local densities. The model adopts an individual-based, spatially explicit approach. In this model, individual sparrows in the population explore a variable landscape consisting of 500- × 500-m cells. This resolution is ecologically appropriate, considering the min-imum territory size, the resolution of many landscape features, and the length of typical "neighborhood" flights.

A set of state variables describes each individual in the population. Individ-uals differ from one another and respond to both the landscape and to other individuals in the population. The minimum set required to model the observed complexity of the behavior of the sparrow includes spatial location, age, sex, weight, reproductive status, fitness, and associations with others. Individual energetics are ignored, it being assumed that if the habitat of a 500- × 500-m cell is an appropriate habitat, individual sparrows will obtain enough food. The model updates the status of each individual daily according to movement and behavior rules. The spine of the model is a simple flow of decisions and actions that affect individuals. At each step the model updates the breeding status and tracks associations between individuals.

Each individual (in random order) moves around the landscape according to a simple set of movement rules. These are dependent upon the time of year, the water levels, the status of the individual, the attributes of the cells it encounters, and the attributes of neighboring cells. Important landscape attributes include elevation, vegetation classification, and fire history. Some types of cells represent "reflective" barriers to movement (pine forest, hammock, and open water); other "transparent" cell types allow movement, but do not represent breeding habitat (sawgrass/spikerush marsh). Temporal and spatial patterns in water levels represent the main environmental driving force behind the model. Males will establish territories when they find an unoccupied area within a spatial cell in which water levels have declined to less than about 5 cm. Nests are built at about 15 cm above ground level and will be abandoned if flooded. A pair of sparrows requires about 45 days to raise a brood.

A set of behavioral rules mimics observed interactions between individuals. The outcome probability of encounters between individuals is dependent upon their relative status. This determines the next movement of each individual, and updates the associations between individuals. For instance, early in the breeding season two neighboring males may fight over the borders of their respective territories. After this stage they reinforce the limits of their territories by countersinging and other less physical behavior. However, males chase neighboring males more often when they are caring for nestlings (Lockwood et al. 1997). Fighting may also be triggered when a bachelor male or juvenile enters an established territory. Normally, the resident male will drive off the intruder.

The direction of unpaired female movements is influenced by the proximity of territorial males. This simulates the fact that male song can be heard (at least by humans) from several hundred meters away. Subsequent encounters between unpaired territorial males and unpaired females may result in successful mating. As breeding activity diminishes the sparrows form small cohesive groups, and associations between individuals become more complex.

SIMSPAR has been used extensively as part of the ATLSS Program to evaluate the impact of hydrological plans on the demographics of the Cape Sable seaside sparrow. These evaluations used a 31-year planning horizon and provided relative assessments of one plan versus another in its impacts on sparrow breeding success, population size, and spatial distribution.

Although this model is simpler than many that will be used in ecosystem management planning, some generalizations can be made from this on the appropriate approach to modeling. First, the model of an ecological system starts with the basic elements, individuals on a dynamically changing landscape. Second, it uses the simplest set of species characteristics essential to the problem of interest: timing of mating behavior and nest initiation, and effects of water levels on initiation and continuation of nesting. Third, it uses relevant information on the primary environmental factor, water level (daily changes in water level in each 500×500 m spatial cell). This model is fairly representative of many of the IBMs used in assessment. It belies the claim of Levin et al. (1997) that "the amount of detail in such models cannot be supported in terms of what we can measure and parameterize..." The IBMs approach is highly advantageous for using the type of data available for specific systems and can be quite parsimonious in its data needs.

Example 2—Wading Birds

A second example is a simulation to evaluate the success of foraging animals over short time periods (as opposed to long time period population models), for which pertinent behavioral information may be the most easily available. Wolff used such a tactical approach for a landscape-level IBM simulating the wood stork (*Mycteria americana*), a wading bird listed as endangered in the U.S. (Wolff 1994; Fleming et al. 1994). This model attempts to predict the breeding success of a wood stork colony under different environmental conditions in the Everglades by simulating the immediate prenesting and nesting periods of these colonial wading birds. Breeding success is a crucial component of the overall health of this population and may be a primary determinant of the viability of the population. It is also readily observable. The individual wood stork forages over a large, heterogeneous landscape, and its success in raising its nestlings depends on the spatial and temporal availability of its food (mainly fish and aquatic macroinvertebrates), which is a strong function of changing water levels within foraging distance of the colony of the individual bird. Wolff developed a model incorporating wading bird bioenergetics and behavioral rules derived from the literature and from discussions with experts on the species. The model makes detailed predictions, based on the foraging capabilities of the wood stork of how different landscape topographies and water management scenarios would alter wood stork reproductive success (Wolff 1994; Fleming et al. 1994). Because reasonably good information is available for all important processes, Wolff's model can make highly specific predictions that should be useful in comparing various possible conservation strategies.

Example 3—Florida Panther/White-Tailed Deer Interaction

The underlying assumption in the model of Wennergren et al.(1995) is a "patch view" of the world, with only two states for any particular patch (suitable and unsuitable), and a view that dispersal mortality is a significant fraction of overall mortality. While such a caricature may be reasonable for some species and habitats, there are many cases for which a spatial continuum of continually varying resources is more appropriate, and in which there is no critical dispersal phase leading to high mortality. This is the case for the third example is an individual-based, spatially explicit model of interacting white-tailed deer and Florida panther populations in South Florida (SIMPDEL, Comiskey et al. 1995).

SIMPDEL (spatially-explicit individual-based simulation model of the Florida panther and white-tailed deer in the Everglades and Big Cypress) includes four major components, hydrology, vegetation, deer, and panthers, and is designed to provide a detailed assessment of how spatial changes in water level affect growth, reproduction, foraging, mating, and predation across South Florida (Comiskey et al. 1998; Abbott et al. 1997; Mellott et al. 1998). It makes use of detailed physiological and behavioral information for the two species, as well as information on vegetative growth under varying hydrologic conditions. Panther movement patterns are derived from radio collar information, and the movements predicted by the model can be explicitly compared to historical movements of individual animals. Data on mortality for deer and panthers have been collected over the past several decades. This allows for realistic levels and causes of mortality to be included, such as deer stranding on high elevation sites during high water conditions, which can lead to starvation, and panther deaths due to intraspecific aggression.

The white-tailed deer, like other large herbivores, forages over a heterogeneous landscape of many localized areas containing resource densities ranging from zero to high levels. This is a case for which a spatial continuum of continually varying resources is more appropriate than the two-state model of Wennergren et al. (1995). This is true as well for large carnivores for which the inherent prey resource, though possibly patchy, moves about in space continually. Such organisms may also have a memory, and elaborate territorial behaviors, which may easily obviate the dispersal error propagation problem the authors infer from their simplified world view. In a continuously distributed resource world, our intuition and model simulations to date do not indicate the strong sensitivity of individual success to small changes in individual movement behavior that the authors claim exists. In this model and others like it (e.g., Hyman et al. 1991; Turner et al. 1995), modeling of populations over many generations seems reasonable.

The above examples illustrate that spatially explicit IBM is actually much broader and more flexible than one would gather from reading the discussion of dispersal in Wennergren et al. (1995). This approach has been developed as a way of taking into account physiological and behavioral processes that could be essential, or at least play a role, in situations involving one or more populations, but that can not be incorporated into the traditional models of population ecology; e.g., small sets of difference or differential equations. The approach makes use of information at the individual organism level that has long been the subject matter of physiological and behavioral ecologists. One can incorporate rules of behavior that are difficult to reduce to simple mathematics.

These examples of IBMs also undermine the pessimistic inference by Wennergren et al. (1995) that IBMs are disadvantageous because they are "data hungry," and the similar criticisms of Levin et al. (1997). For many species of interest, there is a great amount of empirical information already available on behavior and bioenergetics. Rather than being a liability, individual behavior models increase the relevance of behavioral ecology to population ecology. These models are a means for utilizing large amounts of data already collected, often at great cost, at the individual level. The combining of behavioral and physiological information into individual behavior models also helps to reveal gaps in existing data that could stimulate more focused and useful field studies. In many cases, IBMs can already be applied with little or no further demands on data collection, and they can contribute predictive power to conservation problems in a number of ways.

Contrary to the claim by Levin et al. (1997) that IBMs "... represent no real systems," IBMs are clearly being used to address specific questions of specific systems. We believe that for the goal of prediction for specific conservation issues there is no alternative to such detailed site-specific ecological modeling. Abstract ecological models seem to offer little concrete predictive power to conservation ecology. As Shrader–Frechette and McCoy (1993) point out, "... although ecologists' mathematical models may have substantial heuristic power, it may be unrealistic to think that they will ever develop into general laws that are universally applicable and able to provide precise predictions for environmental applications." Generalizations stemming from simple, abstract models are vague, often contradictory, and hotly debated by ecologists (Shrader–Frechette and McCoy 1993). The alleged "shakiness of (detailed) spatial models as a foundation for specific conservation recommendations" cited by Wennergren et al. (1995) should be compared with the questionable foundation for prediction provided by more abstract models.

IBMs and Ecological Theory

The individual-based approach also provides an avenue for important theoretical progress in ecology. E. O. Wilson (1975) forecast that behavioral ecology and population ecology would be tightly interfaced by the end of the 20th century. Much of this interfacing, if it is to occur, will be accomplished through the extension of population models to incorporate the behavior and energetics of individual organisms in a realistic way. This will pave the way towards theory reduction, or interpreting the "higher-level phenomena" of population dynamics in terms of "lower-level processes" or mechanisms at the individual level (Shrader–Frechette and McCoy 1993). Because theory reduction is one of the ultimate goals of science, and because theory reduction is a form of simplification in science, the basing of population modeling on individual behavior is a step toward the consolidation and simplification of ecological theory.

In addition to the impressive empirical work at the individual organism level by behavioral ecologists, there is also highly developed relevant theory at the individual level, such as foraging theory (Stephens and Krebs 1986). If this individual-level theory is judiciously used to help predict energy and time constraints on foraging, the linkages between individual-level theory and population-level theory can be developed.

We disagree with the statement of Levin et al. (1997) that "... only aggregate statistical properties can be reliably predicted, typically over broad spatial and temporal scales." In fact, one can reliably predict that patterns of activity and interaction of individual organisms will lie within bounds imposed by physiological and behavioral constraints at all spatial and temporal scales. This is the whole basis for the use of foraging models and other models of individual animals subject to time and energy constraints.

The wood stork model of Wolff (1994) and white-tailed deer–Florida panther model of Comiskey et al. (1995) are examples of how knowledge of the physiological and behavioral constraints on individuals can be used in models to predict the population-level effects, illustrating theory reduction. Therefore, these models are important not only from an applied viewpoint, but also from a theoretical one. The spatial picture provided by IBMs elucidate the connections between individual-level mechanisms and higher-level patterns, and help to ensure that we are not deceived by superficial resemblances of models to reality at any level of aggregation.

We believe that the development and study of models of this type are essential for understanding the connections between adaptations at the individual level and the dynamics of populations and communities. Models such as the logistic, Lotka-Volterra, McKendrick-von Foerster and the other analytic models of mathematical ecology have served their purposes, but are unable to deal with the fundamental fact that ecological systems are made up of unique individuals in highly complex environments. The desire to produce a unified, parsimonious theory built on the types of equations that have proven so successful in the physical sciences is understandable. But the use of simple analogs of these equations in ecology will go only so far in that direction.

The kinetic equations of physical chemistry, which many models of mathematical ecology emulate, are valid in the domains in which they are used because (1) the basic units (atoms, molecules, or ions) are identical, (2) they are invariant particles that are, for all practical purposes, unchanging, (3) the numbers of these basic units approximate 10^{23}, and (4) approximate spatial uniformity holds in the systems being modeled. None of these facts of physical systems holds for biological populations in natural settings. Each individual in a population differs from all others. A species is not invariant, but has adapted, through natural selection, to its environment. Thus, it is completely dependent on the environment in which it has evolved, down to fine-scale details. Populations of interest are frequently very small, and nearly all populations are too small to justify continuous state variable models such as partial differential equations for describing populations in space.

The hope of many ecological theorists has long been that important ecological problems could be addressed with a few assumptions framed in simple models. This has fostered a style in traditional theoretical ecology of relying on abstract models with no more than a few equations and, therefore, only a few parameters. The use of abstract models has been a successful strategy for generating interesting general theory. But the deficiencies of abstract models are becoming more obvious even in the domain of general theory, because these models cannot incorporate in a realistic way the behavior of organisms, without which ecological theory has only limited applicability to real populations.

For progress to be made in conservation biology and other applied areas of ecology, the traditional abstract models of theoretical ecology are even less likely by themselves to be a successful strategy. The objective of models applied to practical problems should be to bring to bear as much pertinent information on a problem as necessary. This will often include the use of detailed models, when they are supported by data. This is nothing new in the environmental sciences. Environmental scientists and engineers routinely use models with thousands of equations (in hydrology, for example). Wennergren et al. (1995, p. 349), refer to even a modest set of equations for age and spatial structure as "unwieldy," though models of much greater size are hardly termed unwieldy by modelers in other sciences. Whereas Wennergren et al. (1995) state concerning spatially explicit individual behavior models that "...the'realism' of these models is no guarantee of their usefulness," we believe that a high degree of realism is at the very least a prerequisite in any model for it to be useful in conservation ecology. If theoretical ecology is to play a role in conservation and achieve the status of a predictive science, a wide variety of modeling approaches is needed.

While we have focused in this chapter on IBM approaches and argued for their utility in analyzing site-specific environmental problems, the program that these models are a part of takes a broad view of potentially useful approaches. The ATLSS Program (see http://atlss.org/) explicitly includes a multimodeling framework in which a mixture of different modeling approaches are applied. In addition to the IBMs discussed here, ATLSS models include: spatially explicit species index models that produce single values for each spatial cell once a year to summarize the effects of within-year dynamics on the foraging and breeding conditions at a site (Curnutt et al. 1999); and spatially explicit, structured population models that follow the size distribution of populations within each spatial cell (Gaff et al. 1999)

The mixture of approaches in ATLSS allows specification of the organismal, spatial, and temporal level of detail appropriate for the trophic level under consideration and can also account for the limitations imposed by available data. Multiple approaches allow somewhat independent assessments of the impacts of alternative management plans to be made, using different models. As one example of this, predictions of the wading bird model described above may be compared to index models for wading bird breeding potential, which are estimated yearly, and to results from a size-structured fish model that allows within-year tracking of the amount of fish available to wading birds. Conformity of the assessments of management plans drawn from separate models strengthens the utility of such assessments for management. Additionally, using a mixture of models offers the possibility of teasing apart the relative contribution of additional model complexity to the overall assessment.

References

Abbott, C.A., M.W. Berry, E.J. Comiskey, L.J. Gross, and H.-K. Luh. 1997. Computational models of white-tailed deer in the Florida Everglades. *IEEE Computat. Sci. Eng.* **4**:60-72.

Alerstam, T. 1988. *Bird Migration.* Oxford University Press, Oxford, UK.

Allee, W., A. Emerson, O. Park, T. Park, and K. Schmidt, Eds. 1949. *Principles of Animal Ecology.* W. B. Saunders, Philadelphia, PA.

Allen, T.F.H and T.B. Starr. 1982. *Hierarchy: Perspectives for Ecological Complexity.* University of Chicago Press, Chicago, IL.

Alverson, W.W., D.M. Waller, and S.L. Solheim. 1988. Forests too deer: edge effects in northern Wisconsin. *Conserv. Biol.* **2**:348–358.

American Heritage Dictionary of the English Language. 1985. Houghton Mifflin, Boston, MA.

Adrén, H., A. Delin, and A. Seiler. 1997. Population response to landscape changes depends on specialization to different landscape elements, *Oikos* **80**(1): 193–196.

Andrewartha, H.G. 1944. The distribution of plagues of *Austroicetes cruciata* Sauss (Acrididae) in Australia in relation to climate, vegetation and soil. *Trans. R. Soc. South Aust.* **68**:315–326.

Anderson, P.K. 1970. Ecological structure and gene flow in small mammals. *Symp. Zool. Soc. of London* **26**:299–325.

Anderson, R.C. and A.J. Katz. 1994. Recovery of browse sensitive tree species following release from white-tailed deer Odocoileus virginianus Zimmerman browsing pressure. *Biol. Conserv.* **63**:203–208.

Andrén, H. 1994. Effects of habitat fragmentation on birds and mammals in landscapes with different proportions of suitable habitat: a review. *Oikos* **71**:355–366.

Andrén, H. 1996. Population responses to habitat fragmentation: statistical power and the random sample hypothesis. *Oikos* **76**:235–242.

Andrén, H. and A. Delin. 1994. Habitat selection in the Eurasian red squirrel, Sciurus vulgaris, in relation to forest fragmentation. *Oikos* **70**:43–48.

Andrewartha, H.G. and L.C. Birch. 1954. *The Distribution and Abundance of Animals.* University of Chicago Press, Chicago, IL.

Andrewartha, H.G. and L.C. Birch. 1984. *The Ecological Web.* University of Chicago Press, Chicago, IL.

Arts, G.H.P., M. Van Buuren, R.H.G. Jongman, P. Nowicki. D. Wascher, and I.H.S. Hoek. 1995. Editorial. *Landschap,* Special issue on ecological networks. **12**(3):5–9.

Bailey, N.T.J. 1967. The simulation of stochastic epidemics in two dimensions. *Proc. Fifth Berkeley Symp. Math., Stat. Probab.* **4**:237–257.

Baker, W. 1989. A review of models of landscape change. *Landscape Ecol.* **2**(2):111–113.

Balgooyan, C.P. and D.M. Waller. 1995. The use of *Clontonia borealis* and other indicators to gauge impacts of white-tailed deer on plant communities in Northern Wisconsin, USA. *Natl. Areas J.* **15**:308–318.

Ball, G. 1994. The use of GIS in ecosystem modeling. *Environ. Manage.* **18**(3):345–349.

Barbour, T. 1944. *That Vanishing Eden: a Naturalist's Florida.* Little, Brown, Boston, MA.

Bass, O.L., Jr. and J.A. Kushlan. 1982. Status of the Cape Sable sparrow. Report T-672, South Florida Research Center, Everglades National Park. Homestead, FL. 41 pp.

Beecher, W.J. 1955. Late-Pleistocene isolation in salt-marsh sparrows. *Ecology* **36**:23–28.

Behrend, D.F., G.F. Mattfeld, W.C. Tierson, and J.E. Wiley III. 1970. Deer density control for comprehensive forest management. *J. For.* **68**:695–700.

Belsky, A.J. 1995. Spatial and temporal landscape patterns in arid and semi-arid African savannas, in *Mosaic Landscapes and Ecological Processes*. L. Hansson, L. Fahrig, and G. Merriam, Eds. Chapman & Hall, London, 31–56.

Berryman, A.A. 1981. *Population Systems*. Plenum Press, New York.

Berryman, A.A. 1996. What causes population cycles of forest *Lepidoptera? Trends Ecol. Evol.* **11**:28–32.

Birch, L.C. 1957. The role of weather in determining the distribution and abundance of animals, in *Population Studies: Animal Ecology and Demography.* Vol. 22. The Cold Spring Harbor Biological Laboratory, Long Island, NY, 203–215

Bischoff, N.T. and Jongman, R.H.G. 1993. Development of rural areas in Europe: the claim for nature. Netherlands Scientific Council for Government Policy, Preliminary and background studies, V79.

Bissonette. J.A., Ed. 1997. *Wildlife and Landscape Ecology.* Springer-Verlag, New York.

Bissonette, J.A. 1997. Scale-sensitive ecological properties: historical context, current meaning, in *Wildlife and Landscape Ecology.* J.A. Bissonette, Ed. Springer-Verlag, New York, 3–31.

Björnstad, O.N., W. Falck, and N.C. Stenseth. 1995. A geographical gradient in small rodent density fluctuations: a statistical modelling approach. *Proc. R. Soc. London* B **262**:127–133.

Bockstael, N., R. Costanza, I. Strand, W. Boyton, K. Bell, and L. Wagner. 1995. Ecological economic modeling and valuation of ecosystems. *Ecol. Econ.* **14**:143–159.

Bond, W.J. and B.W. van Wilgen. 1996. *Fire and Plants*. Chapman & Hall, New York.

Bormann, F.H. and G.E. Likens. 1979. Pattern and Process in a Forested Ecosystem. Springer-Verlag, New York.

Bosserman, R.W. 1979. The Hierarchical Integrity of *Utricularia*-Periphyton Microecosystems. Ph D. thesis, University of Georgia, Athens, GA.

Botkin, D.B., J.F. Janek, and J.R. Wallis. 1972. Some ecological consequences of a computer model of forest growth. *J. Ecol.* **60**:849–872.

Botkin, D.B., J.M. Mellilo, and L.S.Y. Wu. 1981. How ecosystem processes are linked to large mammal population dynamics, in *Dynamics of Large Mammal Populations.* C.W. Fowler and T.D. Smith, Eds. John Wiley & Sons, New York, 373–387.

Bowring, S.A., D.H. Erwin, Y.G. Jin, M.W. Martin, K. Davidek, and W. Wang 1998. U/Pb ziron geochronology and tempo of the end-Permian mass extinction. *Science* **280**:1039–1045.

Bowyer, R.T., V. Van Ballenberghe, and J.G. Kie. 1997. The role of moose in landscape process: effects of biogeography, population dynamics, and predation, in *Wildlife and Landscape Ecology.* J.A. Bissonette, Ed. Springer-Verlag, New York, 265–287.

Briggs, J. and F.D. Peat. 1984. *Looking Glass Universe: The Emerging Science of Wholeness*. Simon and Schuster, New York.

Brittingham, M.C. and S.A. Temple. 1983. Have cowbirds caused forest song birds to decline? *BioScience.* **33**:31–35.

Brower, L.P. 1995. Understanding and misunderstanding the migration of the monarch butterfly (Nymphalidae) in North America: 1857-1995. *J. Lepid. Soc.* **49**(4):304–385.

Brown, J.H. 1995. *Macroecology.* University of Chicago Press, Chicago, IL.

Brown, J.H. and A.C. Gibson. 1983. *Biogeography.* C.V. Mosby, St. Loius.

Brown, J.H. and A. Kodric-Brown. 1977. Turnover rates in insular biogeography: effects of immigration on extinction. *Ecology* **58**:445–449.

Buçek, A. and J. Lacina. 1992. Territorial system of landscape stability in the CSFR, in Proceedings of the field workshop Ecological Stability of Landscape Ecological Infrastructure Ecological Manage-ment. Federal Committee for the Environment, Institute of Applied Ecology Kostelec n.C.l.

Büdel, J. 1982. *Climate Geomorphology.* Princeton University Press, Princeton, NJ.

Burgess, R.L. and D.M. Sharpe, Eds. 1981. *Forest Island Dynamics in a Man-Dominated Landscape.* Springer-Verlag, New York.

Butler, D.R. 1995. *Zoogeomorphology.* Cambridge University Press, New York.

Cain, M.L. 1985. Random search by herbivorous insects: a simulation model. *Ecology* **66**:876–888.

Callenbach, E. 1996. *Bring Back the Buffalo!* Island Press, Washington, D.C.

Casey, D. and D. Hein. 1983. Effects of heavy browsing on a bird community in deciduous forest. *J. Wild. Manage.* **47**:829–836.

Caswell, H. 1989. *Matrix Population Models.* Sinauer Associates, Sunderland, MA.

Cederlund, G. and R. Bergström. 1996. Trends in the moose-forest system in Fennoscandia, with special reference to Sweden, in *Conservation of Faunal Diversity in Forested Landscapes*, R. DeGraaf and R.I. Miller, Eds. Chapman & Hall, 265-281.

Cheng, B.H.C., R. Bourdeau, and B. Pijanowski. 1996. Systems architecture of an environmental information system incorporating GIS, database management and models in an object-oriented, hierarchical programming design. *J. Photo. Eng. Remote Sensing.*

Cherrett, J.M. 1988. Ecological concepts — the results of the survey of member's views. Bull. B. Ecol. Soc. **19**:80–82.

Clarke, G.L. 1954. Elements of Ecology. John Wiley & Sons, New York.

Clarke, K.C., S. Hoppen, and L. Gaydos. 1997. A self-modifying cellular automation model of historical urbanization in the San Francisco Bay area. *Environ. Plan. Bull.* **24**:247–261.

Clarke, R. 1973. *Ellen Swallow, The Woman Who Founded Ecology.* Follett Publishing Company, Chicago, IL.

Clements, F.E. 1916. Plant succession: An analysis of the development of vegetation. *Carnegie Inst. of Wash.*, Washington, D.C.

Clements, F.E. and V.E. Shelford. 1939. *Bio-ecology.* John Wiley & Sons, New York.

Cohen, J.E. and D. Tilman. 1996. Biosphere 2 and biodiversity: the lessons so far. *Science* **274**:1150–1151.

Collins, S.L., A. Knapp, J.M. Briggs, J.M. Blair, and E.M. Steinauer. 1998. Modulation of diversity by grazing in native tallgrass prairie. *Science* **280**:745–747.

Comiskey, E.J., L.J. Gross, D.M. Fleming, M.A. Huston, O.L. Bass, H.-K. Luh, and Y. Wu. 1995. A spatially explicit individual-based simulation model for Florida panther and white-tailed deer in the Everglades and Big Cypress landscapes. To appear in Florida Panther. Proceedings Volume, U.S. Fish and Wildlife Service.

Comiskey, E.J., L.J. Gross, D.M. Fleming, M.A. Huston, O.L. Bass, H.-K. Luh, and Y. Wu. 1997. A spatially-explicit individual-based simulation model for Florida panther and white-tailed deer in the Everglades and Big Cypress landscapes. *Proceedings of the Florida Panther Conference*, Ft. Myers, Florida, Nov. 1-3, 1994, D. Jordan, Ed., U. S. Fish and Wildlife Service, pp. 494-503.

Connell, J.H. and R.O. Slatyer. 1977. Mechanisms of succession in natural communities and their role in community stability and organization. *Am. Nat.* **111**:1119–1144.

Costanza, R., F. Sklar, and M. White. 1990. Modeling coastal landscape dynamics. *BioScience.* **40**(2):91–107.

Costanza, R., L. Wainger, C. Folke, K.G. Maler. 1993. Modeling complex ecological economic systems. *BioScience.* **43**(8):545–555.

Council of Europe, UNEP and European Centre for Nature Conservation, 1996. The Pan-European Biological and Landscape Diversity Strategy, a vision for Europe's natural heritage.

Croll, I. 1886. *Discussions of Climate & Cosmology.* Appleton Press, New York.

Cronin, T.M. and H.J. Dowsett, Eds. 1991. Pliocene climates. Q. Sci. Rev. **10**:1–282.

Curnutt, J.L., E.J. Comiskey, M.P. Nott, and L.J. Gross. Landscape-based spatially-explicit species index models for Everglades restoration. *Ecolog. Appl.* (in review).

Danell, K. 1977. Dispersal and distribution of the muskrat (*Ondatra zibethica* (L.)) in Sweden. *Viltrevy* **10**:1–26.

Darlington, P.J. Jr. 1957. *Zoogeography.* John Wiley & Sons, New York.

David, J.M., L. Andral, and M. Artois. 1982. Computer simulation of the epi-enzootic disease of vulpine rabies. *Ecol. Model.* **15**:107–125.

Davis, M.B. 1969. Palynology and environmental history during the Quaternary Period. *Am. Sci.* **57**:317–332.

Dawkins, R. 1982. *The Extended Phenotype.* W.H. Freeman, Oxford, UK.

DeAngelis, D.L. and L.J. Gross, Eds. 1992. *Individual-Based Models and Approaches in Ecology: Populations, Communities, and Ecosystems.* Chapman & Hall, New York.

DeAngelis, D.L., B.J. Shuter, M.S. Ridgway, and M. Scheffer. 1993. Modeling growth and survival in an age-0 fish cohort. *Trans. of the Am. Fish. Soc.* **122**:927–942.

DeAngelis, D.L., L.J. Gross, M.A. Huston, W.I. Wolff, D.M. Fleming, E.J. Comiskey, and S.M. Sylveter. 1988. Landscape modeling for Everglades ecosystem restoration. *Ecosystems.* **1**(1):64–74.

DeBach, P. and H.S. Smith. 1941. Are population oscillations inherent in the host-parasite relations? *Ecology* **22**:363–369.

deCalestra, D.S. 1994. Effect of white-tailed deer on songbirds within managed forests in Pennsylvania. *J. of Wild. Manage.* **58**(4):711–217.

deCalestra, D.S. 1995. Deer and diversity in Allegheny hardwood forests: managing an unlikely challenge. *Landscape Urban Plan.* **28**:47–53.

DeGraff, R.M., V.E. Scott, R.H. Hamre, L. Ernst, and S.H. Anderson. 1991. Forest and rangeland birds of the United States. U.S. Department of Agriculture Handbook 688. Washington, D.C.

Delcourt, H.R., P.A. Delcourt, and T. Webb III. 1983. Dynamic plant ecology: the spectrum of vegetational change in space and time. Q. Sci. Rev. **1**:153–175.

den Boer, P.J. 1990. Isolatie en uitsterfkans. De gevolgen van isolatie voor het overleven van populaties van arthropoden. *Landschap* **7**(2):101–120.

den Boer, P.J. 1981. On the survival of populations in a heterogeneous and variable environment. *Oecologia* **50**:39–53.

Denslow, J.S. 1985. Disturbance-mediated coexistence of species, in *The Ecology of Natural Disturbance and Patch Dynamics*. S.T.A. Pickett and P.S. White, Eds. Academic Press, New York. 307–321.

Depew, D.J. and B.H. Weber. 1994. *Darwinism Evolving: Systems Dynamics and the Genealogy of Natural Selection*. MIT Press, Cambridge, MA.

Diamond, H.L. and P.F. Noonan. 1996. *Land Use in America*. Island Press, Washington, D.C.

Diamond, J.M. 1975. The island dilemma: lessons of modern biogeographic studies for the design of natural reserves. *Biol. Conserv.* 7:129–146.

Diamond, J.M. and R.M. May. 1976. Island biogeography and the design of natural reserves, in *Theoretical Ecology: Principles and Applications*. R.M. May, Ed. W. B. Saunders, Philadelphia, PA, 163–186.

Diamond, J.M. and M.E. Gilpin. 1984. Are species co-occurrences on islands non-random, and are null hypotheses useful in community ecology? in *Ecological Communities*. D.R. Strong, D. Simberloff, L.G. Abele, and A.B. Thistle, Eds. Princeton University Press, Princeton, NJ, 297–315.

Diekmann, O., J.A. Metz, and M.W. Sabelis. 1988. Mathematical models of predator/prey/plant interactions in a patch environment. *Exp. Appl. Acarol.* 5(3):319–342.

Dobkin, D.S., I. Olivieri, and P.R. Ehrlich. 1987. Rainfall and the interaction of microclimate with larval resources in the population dynamics of checkerspot butterflies *(Euphydryas editha)* inhabiting serpentine grasslands. *Oecologia* 71:161–166.

Dobson, A.P., J.P. Rodriguez, W.M. Roberts, and D.S. Wilcove. 1997. Geographic distribution of endangered species in the United States. *Science.* 275:550–553.

Dobzhansky, T. 1973. Nothing in biology makes sense except in the light of evolution. *Am. Biol. Teach.* 35:125–129.

Dobzhansky, T., F.J. Ayala, G.L. Stebbins, and J.W. Valentine. 1977. *Evolution*. W.H. Freeman, San Francisco, CA.

Doe, W.W., B. Saghafian, and P. Julien. 1996. Land-use impact on watershed response: the integration of two-dimensional hydrological modeling and geographic information systems. *Hydrol. Proc.* 10:1503–1511.

Donovan, T.M., P.W. Jones, E.M. Annand, and F.T. Thompson III. 1997. Variation in local-scale edge effects: mechanisms and landscape context. *Ecology* 78(7):2064–2075.

Dunning, J.B., B. Danielson, and H.R. Pulliam. 1992. Ecological processes that affect populations in complex landscapes. *Oikos* 65:169–174.

Ehrlich, P.R. 1980. The strategy of conservation, 1980-2000, in *Conservation Biology*. M.E. Soulé and B.A. Wicox, Eds. Sinauer Associates, Sunderland, MA, 329–344.

Ehrlich, P. and A. Ehrlich. 1981. *Extinction*. Random House, New York.

Elzinga, G. and A van Tol. 1994. Groene netwerken voor natuur en recreatie. Otters en natuurgerichte wandelaars, kanoërs en toerfietsers in het Groene Hart. Msc-thesis Wageningen Agricultural University, Department of Physical Planning and Rural Development.

Encyclopedia Americana, International Edition. 1994. Grolier, Danbury, CT.

Engelberg, J. and L.L. Boyarsky. 1979. The noncybernetic nature of ecosystems. *Am. Nat.* 114: 317–324.

Environmental Systems Research Institute. 1996. Cell-based modeling with GRID. Environmental System Research Institute, Atlanta, GA.

Esseen, P.-A., B. Ehnström, L. Ericson, and K. Sjöberg. 1997. Boreal forests. *Ecol. Bull.* 46:16–47.

Fahrig, L. 1983. Habitat Patch Connectivity and Population stability: a model and case study. M.S.c. thesis, Carleton University, Ottawa.

Fahrig, L. and G. Merriam. 1985. Habitat patch connectivity and population survival. *Ecology* 66(6):1762–1768.

Fisher, R.A., A.S. Corbett, and C.B. Williams. 1943. The relation between the number of species and the number of individuals in a random sample of an amimal population. *J. Animal Ecol.* 12:42–58.

Fleming, D.M., W.F. Wolff, and D.L. DeAngelis. 1994. Importance of landscape heterogeneity to wood storks in Florida Everglades. *Environ. Manage.* 18:743–757.

Flintrop, C., B. Hohlmann, T. Jasper, C. Korte, O. Podlaha, S. Scheele, and J. Veizer. 1996. Anatomy of pollution: streams of North Rhine-Wesphalia, Germany. *Am. J. Sci.* 296:58–98.

Forbes, S.A. 1887. The lake as a microcosm. Bulletin of the Peoria Scientific Association, *Ill. Nat. Hist. Surv. Bull.* 15:537–550.

Ford, E.D. and P.J. Diggle. 1981. Competition for light in a plant monoculture modelled as a spatial stochastic process. *Annal. Bot.* 48:481–500.

Forman, R.T.T. 1995. *Land Mosaics. The Ecology of Landscapes and Regions.* Cambridge University Press. Cambridge, UK.

Forman, R.T.T. 1983. Corridors in a landscape: their ecological structure and function. Eko-logia 2:375-387.

Forman, R.T.T., Ed. 1979. *Pine Barrens: Ecosystem and Landscape.* Academic Press, New York.

Forman, R.T.T. 1982. Interaction among landscape elements: a core of landscape ecology, in *Perspectives in Landscape Ecology.* S.P. Tjallingii and A.A. de Veer, Eds. Centre for Agricultural Publishing and Documentation, Wageningen, Netherlands, 35-48.

Forman, R.T.T. and M. Godron. 1986. *Landscape Ecology.* John Wiley & Sons, New York.

Frankel, J.F. and M.E. Soulé. 1981. *Conservation and Evolution.* Cambridge University Press, Cambridge, UK.

Franklin, J.F. 1993. The fundamentals of ecosystem management with applications in the Pacific Northwest, in *Defining Sustainable Forestry.* G.H. Aplet, J.T. Olson, N. Johnson, and V.A. Sample, Eds. Island Press, Washington, D.C., 127–144.

Frelich, L.E. and C.G. Lorimer. 1985. Current and predicted long-term effects of deer browsing in hemlock forests in Michigan, USA. *Biol. Conserv.* 34:99–120.

Frohn, R.C. 1997. *Landscape Ecology.* CRC Press, Boca Raton, FL.

Fuller, T.K. 1989. Population dynamics of wolves in north-central Minnesota. Wild. Mono. 105:1–41.

Fuller, T.K., W.E. Berg, G.L. Radde, M.S. Lenarz, and G.B. Joselyn. 1992. A history and current estimate of wolf distribution and numbers in Minnesota. *Wild. Soc. Bull.* 20:42–55.

Fuller, T.K. 1995. Guidelines for gray wolf management in the northern Great Lakes region. International Wolf Center Technical Publication number 271, International Wolf Center, Ely, MN.

Funderburg, J.B. Jr. and T.L. Quay. 1983. Distributional evolution of the Seaside Sparrow, in *The Seaside Sparrow, Its Biology and Management.* T.L. Quay, J.B. Funderburg, Jr., D.S. Lee, E.F. Porter, and C.S. Robbins, Eds. Occasional Papers of the North Carolina Biological Survey, Raleigh, NC, 19–27.

Gaff, H., D.L. DeAngelis, L.J. Gross, R. Salinas, and M. Shorrosh. A dynamic landscape model for fish in the Everglades and its application to restoration. *Ecological Modelling* (in press).

Gage, D.A., D. Rhodes, K.D. Nolte, W.A. Hicks, T. Leustek, A.J.L. Cooper, and A.D. Hanson. 1997. A new route for synthesis of dimethylsulphoniopropionate in marine algae. *Nature* **387**:891–894.

Garcia-Pera, R. 1994. The Pampas Cat Group (Genus Lynchailurus Severtzov, 1858) (Carnivora: Felidao) A Systematic and Biogeographic Review. American Museum of Natural History.

Gates, J.E. and L.W. Gysel. 1978. Avian nest dispersion and fledging success in field-forest ecotones. *Ecology* **59**(5):871–883.

Gause, G.F. 1934. *The Struggle for Existence*. Williams & Wilkins, Baltimore, 163 pp.

Gause, G.F., N.P. Smaragdova, and A.A. Witt. 1936. Further studies of interaction between predtors and prey. *J. Anim. Ecol.* **5**:1–18.

Gibbons, A. 1997. Y chromosome shows that Adam was an African. *Science* **278**:804-805.

Gibbs, J.W. 1878. On the equilibrium of heterogeneous substances. *Am. J. Sci.* **16**:441–456.

Gilpin, M. and I. Hanski, Eds. 1991. Metapopulation dynamics: empirical and theoretical investigations. *Biol. J. Linn. Soc.* **42**:1–336.

Gilpin, M. and I. Hanski, Eds. 1995. *Metapopulation Dynamics*. Academic Press, London.

Gleason, H.A. 1917. The structure and development of the plant association. *Bull. Torrey Botanical Club.* **44**:463–481.

Gliwicz, J. 1989. Individuals and populations of the bank vole in optimal, suboptimal and insular habitats. *J. Anim. Ecol.* **58**:237–247.

Golley, F.B. 1987. Introducing landscape ecology. *Landscape Ecol.* **1**(1):1–3.

Golley, F.B. 1993. *A History of the Ecosystem Concept in Ecology*. Yale University Press, New Haven, CT.

Golley, F.B., L. Ryszkowski, and J.T. Sokur. 1975. The role of small mammals in temperate forests, grasslands and cultivated fields, *in Small Mammals: Their Productivity and Population Dynamics*. F.B. Golley, K. Petrusewicz, and L. Ryszkowski, Eds. Cambridge University Press, Cambridge, U.K, 223–242.

Goodnight, C.J. 1990a. Experimental studies of community evolution. I. *Evolution* **44**:1614–1624.

Goodnight, C.J. 1990b. Experimental studies of community evolution. II. *Evolution* **44**:1625–1636.

Gould, S.J. and R.F. Johnston. 1972. Geographic variation. *Annu. Rev. Ecol. Syst.* **3**:457–498.

Gould, S.J., Ed. *The Book of Life*. W.W. Norton & Company, New York, 1993.

Graham, R.W. 1986. Response of mammalian communities to environmental changes during the late Quaternary, in *Community Ecology*. J.M. Diamond and T.J. Case, Eds. Harper & Row, New York, 300–313.

Graham, R.W. et al. 1996. Spatial response of mammals to Lake Quaternary environmental fluctuations. *Science.* **272**:1601–1606.

Grumbine, R.E., Ed. 1994a. *Environmental Policy and Biodiversity*. Island Press, Washington, D.C.

Grumbine, R.E. 1994b. What is ecosystem management? *Conserv. Biol.* **8**(1):27–38.

Gunnarsson, B. 1996. Bird predation and vegetation structure affecting spruce-living arthropods in a temperate forest. *J. Anim. Ecol.* **65**:389–397.

Gustafson, E.J. 1998. Quantifying landscape spatial pattern: what is the state of the art? *Ecosystems* **1**:143–156.

Guthrie, R.D. 1970. Bison evolution and zoogeography in North America during the Pleistocene. *Q. Rev. of Biol.* **45**(1):1–15.

Gydemo, R. 1996. Signal crayfish, Pasifastacus leniusculus, as a vector for Psorospermium haeckeli to noble crayfish, Astacus astacus. *Aquaculture* **148**:1–9.

Hagen, J.B. 1992. *An Entangled Bank: the Origins of Ecosystem Ecology.* Rutgers University Press, New Brunswick, NJ.

Haila, Y., O. Järvinen, and R.A. Väisänen. 1979. Effect of mainland population changes on the terrestrial bird fauna of a northern island. *Ornis Scand.* **10**:48–55.

Hamilton, J. 1885. Entomology at Brigantine Beach, NJ in September. *Can. Entomol.* **17**:200–206.

Hammill, J. 1995. Current status of the wolf in Michigan. Abstract. Page 29 *in* Wolves and humans 2000: a global perspective for managing conflict. International Wolf Center, Ely, MN.

Hanski, I. 1994. A practical model of metapopulation dynamics. *J. of Anim. Ecol.* **63**:151–162.

Hanski, I. 1995. Effects of landscape patterns on competitive interactions, in *Mosaic Landscapes and Ecological Processes.* L. Hansson, L. Fahrig, and G. Merriam, Eds. Chapman & Hall, London, 203-224.

Hanski, I. 1996. Habitat destruction and metapopulation dynamics, in *The Ecological Basis of Conservation.* S.T.A. Pickett, R.S. Ostfeld, M. Shachak, and G.E. Likens, editors. Chapman & Hall, London, 217-227.

Hanski, I., L. Hansson, and H. Henttonen. 1991. Specialist predators, generalist predators, and the microtine rodent cycle. *J. Anim. Ecol.* **60**:353–367.

Hanski, I. and H. Henttonen. 1996. Predation on competing rodent species: a simple explanation of complex patterns. *J. Anim. Ecol.* **65**:220–232.

Hanski, I. and M. Gilpin, Eds. 1996. *Metapopulation Dynamics: Ecology, Genetics, and Evolution.* Academic Press, New York.

Hanski, I. and E. Korpimäki. 1995. Microtine rodent dynamics in northern Europe: parameterized model for the predator-prey interactions. *Ecology* **76**:840–850.

Hanski, I., P. Turchin, E. Korpimäki, and H. Henttonen. 1993. Population oscillations of boreal rodents: regulation by mustelid predators leads to chaos. *Nature* **364**:232–235.

Hansson, L. 1971. Small rodent food, feeding and population dynamics. *Oikos* **22**:183–198.

Hansson, L. 1977. Landscape ecology and stability of populations. Landscape Plan. **4**:85–93.

Hansson, L. 1977. Spatial dynamics of field voles *Microtus agrestis* in heterogeneous landscapes. *Oikos* **29**:539–544.

Hansson, L. 1978. Small mammal abundance in relation to environmental variables in three Swedish forest phases. *Stud. Fore. Suec.* **147**:1–40.

Hansson, L. 1983. Bird numbers across edges between mature conifer forest and clearcuts in central Sweden. *Ornis Scand.* **14**:97–103.

Hansson, L. 1989. Landscape and habitat dependence in cyclic and semi-cyclic small rodents. *Holarctic Ecol.* **12**:345–350.

Hansson, L. 1990. Spatial dynamics in fluctuating vole populations. *Oecologia* **85**:213–217.

Hansson, L. 1992. Vole densities and consumption of bark in relation to soil type and bark mineral content. *Scand. J. For. Res.* **7**:229–235.

Hansson, L. 1994. Gradients in herbivory of small mammal communities. *Mammalia* **58**:85–92.

Hansson, L. 1995. Development and application of landscape approaches in mammalian ecology, in *Landscape Approaches in Mammalian Ecology and Conservation.* W.Z. Lidicker, Jr., Ed. University of Minnesota Press, MN, 20-39.

Hansson, L., Ed. 1997. Boreal ecosystems and landscapes: structures, processes and conservation of biodiversity. Ecol. Bull. **46**, 203 pp.

Hansson, L. 1997. Local hot spots and their edge effects: small mammals in oak-hazel woodland. *Oikos,* in press.

Hansson, L. and P. Angelstam. 1991. Landscape ecology as a theoretical basis for nature conservation. Landscape Ecol. **5**(4):191–201.

Hansson, L. L. Fahrig, and G. Merriam, Eds. 1995. *Mosaic Landscapes and Ecological Processes.* Chapman & Hall, London.

Hansson, L. and H. Henttonen. 1985. Gradients in density variations of small rodents: the importance of latitude and snow cover. *Oecologia* **67**:394–402.

Hansson, L. and T.-B. Larsson. 1997. Conservation of boreal environments: a completed research program and a new paradigm, in *Ecological Bulletins* 46. L. Hansson, Ed. Munksgaard International Publishers Ltd. Copenhagen, 9–15.

Hargis, C.D., J.A. Bissonette, and J.L. David. 1997. Understanding measures of landscape pattern, in *Wildlife and Landscape Ecology.* J.A. Bissonette, Ed. Springer-Verlag, New York, New York. 231–261.

Harris, L., T. Hoctor, and S. Gergel. 1996. Landscape processes and their significance to biodiversity conservation, in Population Dynamics in *Ecological Space and Time.* Rodes et al. University of Chicago, Chicago, IL.

Harris, L.D. 1984. *The Fragmented Forest.* University of Chicago Press, Chicago, IL.

Harris, L.D. 1988. Landscape linkages: the dipersal corridor approach to wildlife conservation. *Tran. 53rd North Am. Wild. Nat. Resour. Conf.* **53**:595–607.

Harris, L.D. 1993. Some aspects of biodiversity conservation, in *Our Living Legacy: Proceedings of a Symposium on Biological Diversity.* M.A. Fenger, E.H. Miller, Jr., J.A. Johnson, and E.J.R. Williams, Eds. Royal B C, 97–108.

Harris, L.D. and N.K. Fowler. 1975. Ecosystem analysis and simulation in Mkomazi Reserve, Tanzania. *East Afr. Wild. J.* **13**:325–345.

Harris, L.D. and P.B. Gallagher. 1989. New initiatives for wildlife conservation, in *In Defense of Wildlife: Preserving Communities and Corridors.* G. Mackintosh, Ed. Defenders of Wildlife, Washington, D.C., 11–34.

Harris, L.D., T.S. Hoctor, and S.E. Gergel. 1996. Landscape processes and their significance to biodiversity conservation, in *Population Dynamics in Ecological Space and Time.* O.E. Rhodes, Jr., R.K. Chesser, and M.H. Smith, Eds. University of Chicago Press, Chicago, IL, 319–347.

Harris, L.D. and R. Wallace. 1984. Breeding bird species in Florida forest fragments. *Proc. Annu. Conf. South. Assoc. Fish Wild. Agencies* **83**:87–96.

Harris, L.D. and J. Scheck. 1991. From implications to applications: the dispersal corridor principle applied to the conservation of biological diversity, in *Nature Conservation 2: The Role of Corridors.* D.A. Saunders and R.J. Hobbs, Eds. Surrey Beatty & Sons, Sydney, Australia, 189–220.

Harrison, H. 1984. *Wood Warblers' World.* Simon & Schuster, New York.

Harrison, S. 1991. Local extinction in a metapopulation context: an empirical evaluation. *Biol. J. Linn. Soc.* **42**:73–88.

Harwood, D.M. 1985. Late Neogene climatic fluctuations in the high-southern latitudes: Implications of a warm Gauss and deglaciated Antarctic continent. *South Af. J.Sci.* **81**:239–241.

Hassel, M.P. and R.M. May, Eds. 1990. *Population Regulation and Dynamics.* University Press, Cambridge, UK.

Hastings, J.R. and R.M. Turner. 1980. *The Changing Mile.* The University of Arizona Press. Tucson, AZ.

Hawking, S.W. 1988. *A Brief History of Time: from the Big Bang to Black Holes.* Bantam, New York.

Hawkins, L.K. and P.F. Nicoletto. 1992. Kangaroo rat burrows structure the spatial organization of ground-dwelling animals in a semi-arid grassland. *J. Arid Environ.* **23**:199–208.

Heikkilä, J., A. Below, and I. Hanski. 1994. Synchronous dynamics of microtine rodent populations on islands in Lake Inari in northern Fennoscandia: evidence for regulation by mustelid predators. *Oikos* **70**: 245–252.

Helle, P. 1986. Bird community dynamics in a boreal forest reserve: the importance of large-scale regional trends. *Ann. Zool. Fennici* **23**:157–166.

Hoffmeyer, I. and L. Hansson. 1974. Variability of number and distribution of Apodemus flavicollis (Melch.) and A. sylvaticus in South Sweden. *Z. Säugetierk.* **39**:15–23.

Holling, C.S. 1986. The resilience of terrestrial ecosystems: local surprise and global change, in *Sustainable Development of the Biosphere.* W.C. Clark, and R.E. Munn, Eds. Cambridge University Press, Cambridge, UK, 292–317.

Holling, C.S., D.W. Schindler, B.W. Walker, and J. Roughgarden. 1995. Biodiversity in the functioning of ecosystems: an ecological synthesis, in *Biodiversity Loss: Economic and Ecological Issues.* C. Perrings, K.G. Maler, C. Folke, B.O. Jansson, and C.S. Holling, Eds. Cambridge University Press, New York, 44–83.

Holt, R.D. 1984. Spatial heterogeneity, indirect interactions, and the coexistence of prey species. *Am. Nat.* **124**:377–406.

Hooghiemstra, H. and G. Sarimento. 1991. Long continental pollen record from a tropical intermontane basin: Late Pliocene and Pleistocene history from a 540-meter core. *Episodes* **14**:107–115.

Hooper, D.U. and P.M. Vitousek. 1997. The effects of plant composition and diversity on ecosystem processes. *Science* **277**:1302–1305.

Hoover, J.P. and M.C. Brittingham. 1993. Regional variation in cowbird parasitism of wood thrushes. *Wilson Bull.* **105**(2):228–238.

Hopper, R.G., H.S. Crawford, and R.F. Harlow. 1973. Bird density and diversity as related to vegetation in forest recreation areas. *J. For.* **71**:766–769.

Hörnfeldt, B. 1994. Delayed density dependence as a determinant of vole cycles. *Ecology* 75:791–806.

Hudson, W.E., Ed. 1991. *Landscapes Linkages and Biodiversity.* Island Press, Washington, D.C.

Huffaker, C.B. 1958. Experimental studies on predation: dispersion factors and predator-prey oscillations. *Hilgardia* **27**(14):343–383.

Hunter, M.D. and P.W. Price. 1992. Playing chutes and ladders: heterogenity and the relative roles of bottom-up and top-down forces in natural communities. **73**(3):724–732.

Huston, M.A., D.L. DeAngelis, and W.M. Post. 1988. New computer models may unify ecological theory. *BioScience* **38**:682–691.

Hutchinson, G.E. 1948. Circular causal systems in ecology. *Annu. NY Acad. Scie.* **50**:221–246.

Hyman, J.B., J.B. McAninch, and D.L. DeAngelis. 1991. An individual-based simulation model of herbivory in a heterogeneous landscape, in *Quantitative Methods in Landscape Ecology*. M.G. Turner and R.H. Gardner, Eds. Springer-Verlag, New York, 443–475.

Imbrie, J. and K.P. Imbrie. 1979. *Ice Ages*. Enslow Publishers, Hillside, NJ.

Ims, R.A., J. Rolstad, and P. Wegge. 1993. Predicting space use response to habitat fragmentation: can vole *Microtus oeconomus* serve as an experimental model system (EMS) for capercaillie grouse *Tetrao urogallus* in boreal forests? *Biol. Conserv.* **63**:262–268.

Ims, R.A. and N.C. Stenseth. 1989. Divided the fruitflies fall. *Nature* **342**:21–22.

Jackson, K.T. America's rush to suburbia. *New York Times*, Op-Ed, June 9, 1996.

Janzen, D.H. 1983. No park is an island: increase in interference from outside as park size decreases. *Oikos* **41**:402–410.

Janzen, D.H. 1986. The eternal external threat, in *Conservation Biology*. M.E. Soulé, Ed. Sinauer Associates, Sunderland, MA, 286–303.

Jeffries, M.J. and J.H. Lawton. 1984. Enemy free space and the structure of ecological communities. *Biol. J. Linn. Soc.* **23**:269–286.

Jehoram, S. 1994. Bufferzones waarom en waar om? Naar een toetsingskader voor bufferzoneprojecten. Werkdocument 63 Informatie en Kenniscentrum Natuurbeheer, Wageningen.

Johnston, C.A. 1995. Effects of animals on landscape pattern, in *Mosaic Landscapes and Ecological Processes*. L. Hansson, L. Fahrig, and G. Merriam, Eds. Chapman & Hall, London, 57–80.

Johnston, C.A., J. Pastor, and R.J. Naiman 1993. Effects of beaver and moose on boreal forest landscapes, in *Landscape Ecology and Geographical Information Systems*. S.H. Cousins, R. Haines-Young, and D. Green, Eds. Taylor & Francis, London, 237–254.

Johnston, C.A. and R.J. Naiman. 1990. The use of aeographic information system to analyze long-term landscape alteration by beaver. *Landscape Ecol.* **4**:5–19.

Jones, C.G., J.H. Lawton, and M. Shachak. 1997. Positive and negative effects of organisms as physical ecosystem engineers. *Ecology* **78**(7):1946–1957.

Jones, C.G., J.H. Lawton, and M. Shachak. 1994. Organisms as ecosystem engineers. *Oikos* **69**:373–386.

Jongman, R.H.G. 1995. Ecological networks in Europe, congruent developments. *Landschap* **12**(3):123–130.

Jongman, R.H.G. and Troumbis, A.Y. 1995. The wider Landscape for Nature Conservation: ecological corridors and buffer zones. MN2.7 Project report 1995, submitted to the European Topic Centre for Nature Conservation in Fulfilment of the 1995 Work Programme. European Centre for Nature Conservation.

Jonsson, L. 1992. *Birds of Europe*. Helm, London.

Kaiser, J. 1997. When a habitat is not a home. *Science* **276**:1636–1638.

Karr, J.R. and R.R. Roth. 1971. Vegetation structure and avian diversity in several New World areas. *J. For.* **71**:766–769.

Kavaliauskas, P. 1995. The Nature frame. Lithuanian experience. *Landschap* **12**(3):17–26.

Keast, A. and E. Morton, Eds. 1980. *Migrant Birds in the Neotropics: Ecology, Behavior, Distribution, and Conservation*. Smithsonian Institution Press, Washington, D.C.

Kindvall, O. 1995. The impact of extreme weather on habitat preference and survival in a metapopulation of the bush cricket *Metrioptera bicolor* in Sweden. *Biol. Conserv.* **73**:51–58.

King, A.W. 1997. Hierarchy theory: a guide to system structure for wildlife biologists, in *Wildlife and Landscape Ecology.* J.A. Bissonette, Ed. Springer-Verlag, New York, 185–212.

Kindvall, O. 1996. Habitat heterogeneity and survival in a bush cricket metapopulation. *Ecology* **77**:207–214.

Kolak, J.J., D.T. Long, J.M. Matty, G.J. Larson, and D.F. Sibley. 1996. Groundwater, stream, lake dynamics: Saginaw Bay Watershed. International Association for Great Lakes Research Annual Meeting 1996.

Korpimäki, E., M. Lagerström, and P. Saurola. 1987. Field evidence for nomadism in Tengmalm´s owl *Aegolius funereus. Ornis Scand.* **18**:1–4.

Korpimäki, E., K. Norrdahl, and T. Rinta-Jaskari. 1991. Responses of stoats and least weasels to fluctuating food abundance: is the low phase of the vole cycle due to mustelid predation? *Oecologia* **88**:552–561.

Kuhn, T.S. 1962. *The Structure of Scientific Revolutions.* University of Chicago Press, Chicago, IL.

Kunstler, J.H. 1993. *The Geography of Nowhere.* Touchstone Publishing, New York.

Kushlan, J.A. 1979. Design and management of continental wildlife reserves: lessons from the Everglades. *Biol. Conserv.* **15**:281–290.

Larson, F. 1940. The role of bison in maintaining the short grass plains. *Ecology* **21**:113–121.

Laurance, W.F., S.G. Laurance, L.V. Ferreira, J.M. Rankin-de Merona, C. Gascon, and T.E. Lovejoy. 1997. Biomass collapse in Amazonian forest fragments. *Science* **278**:1117–1118.

Lawton, J.H. 1994. What do species do in ecosystems? *Oikos* **71**:367–374.

Leach, M.K. and T.J. Givnish. 1996. Ecological determinants of species loss in remnant prairies. *Science* **273**:1555–1556.

Leigh, E.F.J. 1991. Genes, bees and ecosystems: the evolution of common interest among individuals. *Trends Ecol. Evol.* **6**:257–262.

Leigh, E.F.J. 1994. Do insect pests promote mutualism among tropical trees? *J. Ecol.* **82**:677–680.

Leopold, A. 1949. *A Sand County Almanac.* Oxford University Press, New York.

Levin, S.A., B. Grenfall, A. Hastings, and A.S. Perelson. 1997. Mathematical and computational challenges in population biology and ecosystem science. *Science* **275**:334–343.

Levins, R. 1969. Some demographic and genetic consequences of environmental heterogeneity for biological control. *Bull. Entomol. Soc. Am.* **15**:237–240.

Levins, R. 1970. Extinction, in *Some Mathematical Questions in Biology.* M. Gerstenhaber, Ed. American Mathematical Society, Providence, RI, 77–107.

Lewin, R. 1984. Why is development so illogical? *Science.* **224**:1327–1329.

Lidicker, W.Z., Jr. 1995. The landscape concept: something old, something new, in *Landscape Approaches in Mammalian Ecology.* W.Z. Lidicker, Jr., Ed. University of Minnesota Press, Minneapolis, MN, 3–19.

Likens, G.E., C.T. Driscoll, and D.C. Buso. 1996. Long-term effects of acid rain: response and recovery of a forest ecosystem. *Science.* **272**:244–246.

Lindeman, R.L. 1942. The trophic-dynamic aspect of ecology. *Ecology* **23**(4):399–418.

Lindström, E.R. and B. Hörnfeldt. 1994. Vole cycles, snow depth and fox predation. *Oikos* **70**:156–160.

Lockwood, J.L., R.D. Powell, M.P. Nott, and S.L. Pimm. 1997. Assembling ecological communities in time and space. *Oikos.* **80**(3): 549–553.

Long, J.L. 1981. *Introduced Birds of the World.* David & Charles, London.

Lovejoy, T.E., R.O. Bierregaard, J.M. Rankin, and H.O.R. Schubart. 1983. Ecological dynamics of tropical forest fragments, in *Tropical Rain Forest: Ecology and Management*. S.L. Sutton, T.C. Whitmore, and A.C. Chadwick, Eds. Blackwell, Oxford, 377–384.

Lovejoy, T.E., J.M. Rankin, R.O. Bierregaard, K.S. Brown, L.H. Emmons, and M.E. Van der Voort. 1984. Ecosystem decay of Amazon forest fragments, in *Extinctions*. M.H. Nitecki, Ed. University of Chicago Press, Chicago, IL, 295–325.

Lovelock, J.E. 1979. Gaia: A New Look at Life on Earth. Oxford University Press, NY.

Lyell, C. Sir. 1830–1833. *Principles of Geology*. John Murray, London.

MacArthur, R.H. and J.W. MacArthur. 1961. On bird species diversity. *Ecology* **42**:594–598.

MacArthur, R.H. and E.R. Pianka. 1966. On optimal use of a patchy environment. *Am. Nat.* **100**:603–609.

MacArthur, R.H. and E.O. Wilson. 1963. An equilibrium theory of insular zoogeography. *Evolution.* **17**:373–387.

MacArthur, R.H. and E.O. Wilson. 1967. *The Theory of Island Biogeography*. Princeton University Press, Princeton, NJ.

MacCluer, J.W. 1967. Monte Carlo methods in human population genetic: a computer model incorporating age-specific birth and death rates. *Am. J. Hum. Genet.* **19**:303–312.

Maehr, D. 1996. *The Florida Panther: Life and Death of a Vanishing Carnivore*. Island Press, Washington, D.C.

Maguire, L.A. and J.W. Porter. 1977. A spatial model of growth and competition strategies in coral communities. *Ecol. Model.* **3**:249–271.

Maidment, D.R. 1991. GIS and hydrologic modeling. Proceedings of the First International Symposium on GIS and Environmental Modeling, Boulder, CO.

Malakoff, D. 1998. Restored wetlands flunk real-world test. *Science* **280**:371–372.

Malin, G. 1997. Sulphur, climate and the microbial maze. *Nature* **387**:857–859.

Malthus, T.R. 1798. *An Essay on the Principle of Population as it Affects the Future Improvement of Society*. J. Johnson, London, UK.

Mander, Ü., J. Jagomägi, and M. Külvik. 1989. Network of compensative areas as an ecological infrastructure of territories, in *Connectivity in Landscape Ecology*. Proceedings of the 2nd International Seminar of the International Association for Landscape Ecology, Münster (1987). K-F Schreiber, Ed. Münstersche Geographische Arbeiten 29, pp. 35–38.

Markgraf, V., E. Romero, and C. Villagrán. 1996. History and paleoecology of South American *Nothofagus* forests, in *The Ecology and Biogeography of Nothofagus Forests*. T.T. Veblen, R.S. Hill, and J. Read, Eds. Yale University Press, New Haven, CT, 354–386.

Marquis, R.J. and C.J. Whelan. 1994. Insectivorous birds increase growth of white oak through consumption of leaf-chewing insects. *Ecology.* **75**(7):2007–2014.

Marsh, G.P. 1907. *The Earth as Modified by Humans*. Charles Schribner's Sons, New York.

May, R.M. and S.K. Robinson. 1985. Population dynamics of avian brood parasitism. *Am. Nat.* **126**:475–494.

Mayr, E. 1963. *Animal Species and Evolution*. Harvard University Press, Cambridge, MA.

McDonald, J.N. 1981. *North American Bison*. University of California Press, Berkeley, CA.

McGarrahan, E. 1997. Much-studied butterfly winks out on Stanford preserve. *Science* **275**:479–480.

McLaren, B.E. and R.O. Peterson. 1994. Wolves, moose, and tree rings on Isle Royale. *Science.* **226**:1555–1558.

McNaughton, S.J., F.F. Banyikwa, and M.M. McNaughton. 1997. Promotion of the cycling of diet-enhancing nutrients by African grazers. *Science* **278**:1978–1800.

McShea, W.J. and J.H. Rappole. 1992. White-tailed deer as keystone species within forested habitats of Virginia. *Va. J. Sci.* **43**:177–186.

Mech, L.D. 1995. The challenge and opportunity of recovering wolf populations. *Conserv. Biol.* **9**:270–278.

Mech, L.D., S.H. Fritts, G.L. Radde, and W.J. Paul. 1988. Wolf distribution and road density in Minnesota. *Wild. Soc. Bull.* **16**:85–87.

Mech, L.D., S.H. Fritts, and D. Wagmer. 1995. Minnesota Wolf dispersal to Wisconsin and Michigan. *Am. Midland Naturalist.* **133**:368–370.

Mech, L.D. and R.M. Nowak. 1981. Return of the grey wolf to Wisconsin. *Am. Mid. Nat.* **105**:408–409.

Mellott, L.E., M.W. Berry, E.J. Comiskey, and L.J. Gross. 1999. The design and implementation of an individual-based predator-prey model for a distributed computing environment. *Simulation Practice and Theory* **7**:47-70.

Melquist, W. and M. Hornocker. 1983. Ecology of river otters in west central Idaho. *Wild. Monog.* 83. 60 pp.

Mercer, J.H. 1987. The Antarctic ice sheet during the late Neogene, in *Palaeocology of Africa and the Surrounding Islands.* J.A. Coetzee, Ed. Balkema, Rotterdam, 21–33.

Merriam, G. 1988. Landscape ecology: the ecology of heterogeneous systems, in *Landscape Ecology and Management.* M.R. Moss, Ed. Polyscience Publications, Montreal, 43–50.

Merriam, G. 1991. Corridors and connectivity: animal populations in heterogeneous environments, in *Nature Conservation 2: the Role of Corridors.* D.A. Saunders and R.J. Hobbs, Eds. Surrey Beatty & Sons, New South Wales, Australia, 133-142.

Merriam, H.G. 1984. Connectivity: a fundamental ecological characteristic of landscape pattern, in *Proceedings of the First International Seminar on Methodology in Landscape Ecological Research and Planning.* Roskilde, Denmark. International Association for Landscape Ecology, 5–15.

Merriam, G. and J. Wegner. 1992. Local extinction, habitat fragmentation and ecotones, in *Landscape Boundaries: Consequences for Biodiversity and Ecological Flows.* A.J. Hansen and F. di Castri, Eds. Springer-Verlag, New York, 150-169.

Michigan Department of Natural Resources. 1994. *Saginaw Bay Watershed Prioritization Process.* MDNR. 65 pp.

Michigan United Conservation Clubs. 1993. *Saginaw Bay Watershed Land Use and Zoning Study.* 308 pp.

Middleton, J. and G. Merriam. 1981. Woodland mice in a farmland mosaic. *J. Appl. Ecol.* **18**:703–710.

Middleton, J. and G. Merriam. 1983. Distribution of woodland species in farmland woods. *J. Appl. Ecol.* **20**:625–644.

Miklós, L. 1996. The concept of the territorial system of landscape stability in Slovakia, in Ecological and landscape consequences of land use change in Europe. Proceedings of the first ECNC seminar on land use change and its ecological consequences. R. Jong-man, Ed. ECNC Publication series on man and nature Vol. 2, pp. 385–406.

Miller, R.I. and L.D. Harris. 1977. Isolation and extirpation in wildlife reserves. *Biol. Conserv.* **12**:311–315.

Ministry of Agriculture, Nature Management and Fisheries 1990. Nature Policy Plan of the Netherlands. The Hague.

Ministry of Transport, Public Works and Water Management. 1995. Wildlife crossings for roads and waterways. Delft.

Mladenoff, D.J. and F. Sterns. 1993. Eastern hemlock regeneration and deer browsing in the Northern Great Lakes Region: a re-examination and model simulation. *Conserv. Biol.* **7**:889–900.

Mladenoff, D.J., M.A. White, T.R. Crow, and J. Pastor. 1994. Applying principles of landscape design and management to integrate old-growth forest enhancement and commodity use. *Conserv. Biol.* **8**:752–762.

Mladenoff, D.J., T.A. Sickley, R.G. Haight, and A.P. Wydeven. 1995. A regional landscape analysis and prediction of favorable gray wolf habitat in the Northern Great Lakes Region. *Conserv. Biol.* **9**:279–294.

Mladenoff, D.J., R.G. Haight, T.A. Sickley, and A.P. Wydeven. 1997. Causes and implications of species restoration in altered ecosystems. *BioScience* **47**(1):21–31.

Mooij, W.M. and D.L. DeAngelis. 1999. Error propagation in spatially explicit population models: A reassessment. *Conserv. Biol.* **13**:930-933.

Mun, H.-T. and W.G. Whitford. 1990. Factors affecting annual plant assemblages on banner-tailed kangaroo rat mounds. *J. Arid Environ.* **18**:165–173.

Naiman, R.J. 1988. Animal influences on ecosystem dynamics. *BioScience* **38**(11):750–752.

Naiman, R.J., C.A. Johnston, and J.C. Kelley. 1988. Alteration of North American streams by beaver. *BioScience* **38**(11):753–762.

Naveh, Z. 1991. Some remarks on recent developments in landscape ecology as a transdisciplinary ecological and geographical science. *Landscape Ecol.* **5**(2):65–73.

Naveh, Z. 1982. Landscape ecology as an emerging branch of human ecosystem science, in *Advances in Ecological Research*, Vol. 12. A. MacFadyen and E.D. Ford, Eds. Academic Press, New York, 189–237.

Naveh, Z. and A.S. Lieberman. 1984. *Landscape Ecology: Theory and Application.* Springer-Verlag, New York.

Neff, E. 1967. *Die theoretischen grundlagen landschaftslehre.* Verlag VEB Herman Harck, Geogrophisch-Kartogrophische Anstalt Gotha, Leipzig.

Netherlands Scientific Council for Government Policy. 1992. Ground for Choices. Four perspectives for the rural areas in the European Community. Reports to the Government 42. The Hague, SDU uitgeverij.

Newmark, W.D. 1987. A land-bridge island perspective on mammalian extinctions in western North American parks. *Nature* **325**(29):430–432.

Newmark, W.D. 1995. Extinction of mammal populations in western North American National Parks. *Conserv. Biol.* **9**(3):512–526.

Newsome, A.E., P.C. Catling, and L.K. Corbett. 1983. The feeding ecology of the dingo. II. Dietary and numerical relationships with fluctuating prey populations in southeastern Australia. *Aust. J. Ecol.* **8**:345–366.

Nicholson, A.J. 1933. The balance of animal populations. *J. Anim. Ecol.* **2**, Suppl. 132–178.

Nicholson, A.J. 1954. An outline of the dynamics of animal populations. *Aust. J. Zoo.* **2**:9–65.

Nilsson, S.G. 1979. Density and species richness of some forest bird communities in South Sweden. *Oikos* **33**:392–401.

Norton, O.R. 1994. *Rocks from Space*. Mountain Press Publishing, Missoula, MT.

Noss, R.F. 1983. A regional landscape approach to maintain diversity. *BioScience* **33**(11):700–706.

Noss, R.F. 1993. Sustainable forestry on sustainable forests, in *Defining Sustainable Forestry*. G.A. Aplet, J.T. Olson, N. Johnson, and V.A. Sample, Eds. Island Press, Washington, D.C., 12–42

Noss, R.F. and L.D. Harris. 1986. Nodes, networks, and MUMs: preserving diversity at all scales. *Environ. Manage.* **10**(3):299–309.

Nott, M.P. and D. DeAngelis. 1997. Modeling the effects of hydrological regimes on the distribution and breeding success of the Cape Sable seaside sparrow in the Florida Everglades. *Bull Ecol. Soc. Am.* **78**(4):26

Nummi, P. and H. Pöysä. 1997. Population and community level responses in Anas-species to patch disturbance caused by an ecoststem engineer, the beaver. *Ecography* **20**:580–584

Odum, E. *Fundamentals of Ecology*, 3rd ed. W.B. Saunders Company, Philadelphia,PA, 8.

Odum, E.P. 1959. *Fundamentals of Ecology*, 2nd ed. Sanders, Philadelphia, PA.

Odum, E.P. 1969. The strategy of ecosystem development. *Science* **164**:262–270

Odum, 1971. Environment, Power and Society. Wiley. Interchange, New York.

Odum, H.T. 1997. *Environment, Power and Society*. Wiley, John & Sons, New York.

Ojima, D., K. Galvin, and B.L. Turner. 1994. The global impact of land-use change. *BioScience* **44**(3):300–304.

Okubo, A. 1980. *Diffusion and Ecological Problems: Mathematical Models*. Springer-Verlag, Berlin.

Oldfield, M. 1989. *The Value of Conserving Genetic Resources*. Sinaur Associates. Sunderland, MA.

Oldfield, S. 1988. Buffer zone management in tropical moist forest: case studies and guidelines. IUCN, Gland, Switzerland.

O'Neill, R.V., D.L. DeAngelis, J.B. Waide, and T.F.H. Allen. 1986. *A Hierarchical Concept of the Ecosystem*. Princteon University Press, Princeton, NJ.

O'Neill, R.V., A.R. Johnson, and A.W. King. 1989. A hierarchical framework for the analysis of scale. *Landscape Ecol.* **3**:193–205.

Opdam, P.F.M. 1991. Metapopulation theory and habitat fragmentation: a review of holarctic breeding bird studies. *Landscape Ecol.* **5**(2):93–106.

Orr, R.T. 1970. *Animals in Migration*. MacMillan, New York.

Pastor, J., R.J. Naiman, B. Dewey, and P. McInnes. 1988. Moose, microbes, and the boreal forest. *BioScience* **38**:770–777.

Pastor, J., B. Dewey, R. Moen, D.J. Mladenoff, M. White, and Y. Cohen. 1998. Spatial patterns in the moose-forest-soil ecosystem on Isle Royale, Michigan, USA. *Ecol. Appl.* **8**(2):411–424.

Peters, R.H. 1991. *A Critique for Ecology*. Cambridge University Press, Cambridge, MA.

Peterson, R.O. 1988. The pit or the pendulum: issues in large carnivore management in natural ecosystems, in *Ecosystem Management for Parks and Wilderness*. J.K. Agee and D.R. Johnson, Eds. University of Washington Press, Seattle, Washington, 105–117.

Pettersson, R., J.P. Ball, K.E. Renhorn, P.-A. Esseen, and K. Sjöberg. 1995. Invertebrate communities in boreal forest canopies as influenced by forestry and lichens with implications for passerine birds. *Biol. Conserv.* **74**:67–73

Pickett, S.T.A. and P.S. White. 1985. Patch dynamics: a synthesis, in *The Ecology of Natural Disturbance and Patch Dynamics*. S.T.A. Pickett and P.S. White, Eds. Academic Press, New York, 371–383.

Pickett, S.T.A., J. Kolasa, and C.G. Jones. 1994. *Ecological Understanding*. Academic Press, New York.

Pielou, E.C. 1991. *After the Ice Age*. University of Chicago Press, Chicago, IL.

Pienaar, U.d.V. 1968. The ecological significance of roads in a nature park. *Koedoe* 11:169–175.

Pijanowski, B.C., T. Machemer, S. Gage, D. Long, W. Cooper, and T. Edens. 1995. A land transformation model: integration of policy, socioeconomics and ecological succession to examine pollution patterns in watershed. Report to the Environmental Protection Agency, Research Triangle Park, N.C.

Pijanowski, B.C., T. Machemer, S. Gage, D. Long, W. Cooper, and T. Edens. 1996. The use of a geographic information system to model land use change in the Saginaw Bay Watershed. Proceedings of the Third International Conference on GIS and Environmental Modeling, Sante Fe, New Mexico, January 21–26, 1996. GIS World Publishers on CD-ROM.

Pijanowski, B.C., T. Machemer, S. Gage, and D. Long. In review. A land transformation model: conceptual and analytical framework and its application to Michigan's Saginaw Bay Watershed. To Ecological Modeling.

Pimm, S.L. and A. Redfern. 1988. The variability of animal populations. *Nature* 334:613–614.

Ponting, C. 1991. *A Green History of the World: The Environment and the Collapse of Great Civilizations*. Sinclair-Stevenson Limited, London.

Popper, K.R. 1959. *The Open Universe: an Argument for Indeterminism*. Rowman and Littlefield, Totowa, NJ.

Popper, K.R. 1990. *A World of Propensities*. Thoemmes, Bristol, UK.

Post, W. and J.S. Greenlaw. 1994. Seaside Sparrow, in *The Birds of North America*, No. 127. A. Poole and F. Gill, Eds. The Academy of Natural Sciences of Philadelphia.

Potts, R. 1997. *Humanity's Descent*. Avon Books, New York.

Powers, S. 1911. Floating islands. *Pop. Sci. Mon.* 79:303–307.

Preston, F.W. 1960. Time and space and the variation of species. *Ecology* 29:254–283.

Preston, F.W. 1962. The cononical distribution of commonness and rarity: Part I. *Ecology.* 43:185–215.

Preston, F.W. 1962. The cononical distribution of commonness andrarity: Part II. *Ecology.* 43:410Δ–432.

Pucek, Z., W. Jedrzejewski, B. Jedrzejewska, and M. Pucek. 1993. Rodent population dynamics in a primeavel deciduous forest (Bialowieza National Park) in relation to weather, seed crop, and predation. *Acta Theriol.* 38:199–232.

Pulliam, H.R. 1988. Sources, sinks, and population regulation. *Am. Nat.* 132(5):652–661.

Ream, R.R., M.W. Fairchild, D.K. Boyd, and D.H. Pletscher. 1991. Population dynamics and home range changes in a colonizing wold population, in *The Greater Yellowstone Ecosystem: Redefining America's Wilderness Heritage*. R.B. Keiter and M.S. Boyce, Eds. Yale Universtiy Press, New Haven, CT.

Reijnen, R., W.B. Harms, R.P.B. Foppen, R. de Visser, and H.P. Wolfert. 1995. Rhine Econet, Ecological networks in river rehabilitation scenarios: a case study for the lower Rhine. Publications and reports of the project Ecological Rehabilitation of the Rivers Rhine and Meuse, no58. Riza Lelystad.

Reynolds, J. C. 1985. Details of the geographic replacement of the red squirrel *(Sciurus vulgaris)* by the grey squirrel *(Sciurus carolinensis)* in eastern England. *J. Anim. Ecol.* **54**:149–162.

Richards, C.G.J. 1985. The population dynamics of *Microtus agrestis* in Wytham, 1949 to 1978. *Acta Zool. Fenn.* **173**:35–38.

Riebsame, W., W. Parton, K. Galvin, I. Burke, L. Bohren, R. Young, and E. Knop. 1994. Integrated modeling of land use and cover change. *BioScience* **44**(5):350–356.

Richards, C., L. Johnson, and G. Host. 1993. Landscape influences on habitat, water chemistry and macroinvertebrate assemblages in Midwestern stream ecosystems. Environmental Research Laboratory, U.S. EPA, Duluth, MN 55804. NRRI Technical Report No. TR-93-109.

Ridley, M. 1993. *Evolution*. Blackwell Scientific, Cambridge, MA.

Risser, P.G., J.R. Karr, and R.T.T. Forman. 1984. *Landscape Ecology*. Illinois Natural History Survey Special Publication Number 2, Champaign, IL.

Robinson, S.K. 1993. Conservation problems of Neotropical migrant land birds. *Trans. North Am. Wild. Nat. Resour. Conf.* **58**:379–389.

Robinson, S.K., F.R. Thompson, III, T.M. Donovan, D.R. Whitehead, and J. Faaborg. 1995. Regional forest fragmentation and the nesting success of migratory birds. *Science* **267**:1987–1990.

Rohlf, F.J. and D. Davenport. 1969. Simulation of simple models of animal behavior with a digital computer. *J. Theoret. Biol.* **23**:400–424.

Rosen, R. 1985. Information and complexity, in *Ecosystem Theory for Biological Oceanography*. R.E. Ulanowicz and T. Platt, Eds. Canadian Bulletin of Fisheries and Aquatic Sciences 213, Ottawa, 221–233.

Rotar, J.P. and M. Adamic. 1997. Wildlife-traffic relations in Slovenia, in Proceedings of the international conference on habitat fragmentation, infrastructure and the role of ecological engineering, 17–21 September 1995, Maastricht-The Hague. K.Canyers, Ed. Ministry of Transport, Public Works and Water Management. Habitat fragmentation and Infrastructure, pp. 86–92.

Rowe, J.S. 1988. Landscape ecology: the study of terrain ecosystems, in *Landscape Ecology and Management*. M.R. Moss, Ed. Polyscience Publications, Montreal, 35–42.

Ruckelshaus, M., C. Hartway, and P. Kareiva. 1997. Assessing the data requirements of spatially explicit dispersal models. *Conserv. Biol.* **11**:1298-1306.

Rütimeyer, K.L. 1869. Über Thal — und Seebildung. Schweighauser. Basel, Switzerland.

Rykiel, E.J., Jr. 1996. Testing ecological models: the meaning of validation. *Ecol. Model.* **90**:228–244.

Salthe, S.N. 1993. *Development and Evolution: Complexity and Change in Biology*. MIT Press, Cambridge, MA.

Scheiner, S.M., A.J. Hudson, and M.A. VanderMeulen. 1993. An epistemology for ecology. *Bull. Ecol. Soc. Am.* **74**:17–21.

Schneider, S.H. and R. Londer. 1984. *The Coevolution of Climate and Life*. Sierra Club Books, San Francisco, CA.

Schonewald-Cox, C.M., S.M. Chambers, B. Macbryde, and L. Thomas, Eds. 1983. *Genetics and Conservation: a Reference for Managing Wild Animal and Plant Populations*. Benjamin/Cummings, Menlo Park, CA.

Schull, W.J. and B.R. Levin. 1964. Monte Carlo simulation: some uses in the genetic study of primitive man, in *Stochastic Models in Medicine and Biology*. J. Gurland, Ed. University of Wisconsin Press, Madison, WI, 179–196.

Schultz, J. 1995. *Ecosystems of the World*. Springer-Verlag, New York.

Shafer, C. 1994. Beyond park boundaries, in *Landscape Planning and Ecological Networks*. E.A. Cook and H.N. van Lier, Eds. Elsevier, New York, 201–223.

Shaffer, M.L. 1981. Minimum populations sizes for species conservation. *BioScience* **31**:131–134.

Shilling, F. 1997. Do habitat conservation plans protect endangered species? *Science* **276**:1662–1663.

Shannon, H.J. 1916. Insect migration as related to those of birds. *Sci. Mon.* **1916**:227–240.

Shrader-Frechette, K.S. and E.D. McCoy. 1993. *Method in Ecology*. Cambridge University Press, Cambridge, UK.

Simberloff, D. 1998. Flagships, umbrellas, and keystones: is single-species management passé in the landscape era? *Biol. Conserv.* **83**(3):247–257.

Simberloff, D. 1980. A succession of paradigms in ecology: essentialism to materialism and probabilism. *Synthese* **43**:3–39.

Simberloff, D.S. and L.G. Abele. 1976. Island biogeography and conservation practice. *Science* **191**:285–286.

Simberloff, D. and J. Cox. 1987. Consequences and costs of conservation corridors. *Conserv. Biol.* **1**:63–71.

Simberloff, D., J.A. Farr, J. Cox, and D.W. Mehlman. 1992. Movement corridors: conservation bargains or poor investments. *Conserv. Biol.* **6**:493–504.

Simpson, G.G. 1936. Data on the relationships of local and continental mammalian faunas. *J. Palaeontol.* **10**: 410–414.

Simpson, G.G. 1940. Mammals and land bridges. *J. Wash. Acad. of Sci.* **30**:137–163.

Siniff, D.B. and C.R. Jessen. 1969. A simulation model of animal movement patterns. *Adv. Ecol. Res.* **6**:185–217.

Sjögren Gulve, P. 1994. Distribution and extinction patterns within a northern metapopulation of the pool frog, *Rana lessonae. Ecology* **75**:1357–1367.

Sjögren, P. 1991. Extinction and isolation gradients in metapopulations: the case of the pool frog *(Rana lessonae). Biol. J. Linn. Soc.* **42**:135–147.

Sjögren Gulve, P. and C. Ray. 1996. Using logistic regression to model metapopulation dynamics: Large-scale forestry extirpates the pool frog, in *Metapopulations and Wildlife Conservation*. D. McCullough, Ed. Island Press, Washington, D.C., 111–137.

Skellam, J.G. 1973. The formulation and interpretation of mathematical models of diffusionary processes in population biology, in *The Mathematical Theory of the Dynamics of Biological Populations*. M.S. Bartlett and R.W. Hiorns, Eds. Academic Press, New York, 63–85.

Sklar, F. and R. Costanza. 1990. The development of dynamic spatial models for landscape ecology: a review and prognosis, in *Quantitative Methods in Landscape Ecology: The Analysis and Interpretation of Landscape Heterogeneity*. Springer-Verlag, New York.

Slatkin, M. 1987. Gene flow and the geographic structure of natural populations. *Science* **236**:787–792.

Smithsonian Institution. 1994. The science of overabundance: the ecology of unmanaged deer populations. Proceedings Smithsonian Institution. Washington, D.C.

Soulé, M.E., Ed. 1987. *Viable Populations for Conservation*. Cambridge University Press, Cambridge, UK.

Soulé, M.E. and B.A. Wilcox, Eds. 1980. *Conservation Biology*. Sinauer Associates, Sunderland, MA.

Stamps, J.A., M. Buechner, and V.V. Krishnan. 1987. The effects of edge permeability and habitat geometry on emigration from patches of habitat. *Am. Nat.* **129**:533–552.

Stehli, F. and S.D. Webb, Eds. 1985. *The Great American Biotic Interchange.* Plenum Press, New York.

Stenseth, N.C. 1980. Spatial heterogeneity and population stability: some evolutionary consequences. *Oikos* **35**:165–184.

Stenseth, N.C., Ed. 1993. Academic Press, New York.

Stenseth, N.C. and W.Z. Lidicker. 1992. The use of radioisotopes in the study of dispersal: with a case study, in *Animal Dispersal: Small Mammals as a Model.* N.C. Stenseth and W.Z. Lidicker, Eds. Chapman & Hall, London, 333–352.

Stephens, D.W. and J.R. Krebs. 1986. *Foraging Theory.* Princeton University Press, Princeton, NJ.

Sutherland, W.J. 1996. *From Individual Behaviour to Population Ecology.* Oxford University Press, Oxford, UK.

Swanson, F.J., T.K. Kratz, N. Caine, and R.G. Woodmansee. 1988. Landform effects on ecosystem patterns and processes. *BioScience* **38**(2):92–98.

Tansley, A.G. 1935. The use and abuse of vegetational concepts and terms. *Ecology* **16**(3):284–307.

Tansley, A.G. 1939. *The British Isles and Their Vegetation.* Cambridge University Press, Cambridge, UK.

Taylor, P.D., L. Fahrig, K. Henein, and G. Merriam. 1993. Connectivity is a vital element of landscape structure. *Oikos* **68**:571–573.

Temple, S.A. and J.R. Cary. 1988. Modelling dynamics of habitat-interior bird populations in fragmented landscapes. *Conserv. Biol.* **2**:340–347.

Tenow, O. and A. Nilssen. 1990. Egg cold hardiness and topoclimatic limitations to outbreaks of Epirrita *autumnata* in northern Fennoscandia. *J. Appl. Ecol.* **27**:723–734.

Terborgh, J. 1976. Island biogeography and conservation: strategy and limitations. *Science* **193**(4257):1029–1030.

Terborgh, J. 1989. *Where Have All the Birds Gone?* Princeton University Press, Princeton, NJ.

Thompson, W.A., I. Vertinsky, and J.R. Krebs. 1974. The survival value of flocking in birds: a simulation model. *J. Anim.Ecol.* **43**:785–808.

Tilghman, N.G. 1989. Impacts of white-tailed deer on forest regeneration in northwestern Pennsylvania. *J. Wild. Manage.* **53**:524–532.

Tilman, D. and P. Kareiva. 1997. *Spatial Ecology.* Princeton University Press, Princeton, NJ.

Tjallingii, S.P. and A.A. de Veer, Eds. 1982. *Perspectives in Landscape Ecology.* Centre for Agricultural Publishing and Documentation, Wageningen, Netherlands.

Tribus, M. and E.C. McIrvine. 1971. Energy and information. *Sci. Am.* **225**:179–188.

Tricart, J. 1965. *Principes et Méthodes de la Géomorphologie.* Masson, Paris.

Troll, C. 1968. Landschaftsokogie, in *Pfanzensoziologic und landscahftsokogie.* R. Tuxen, Ed. Verlag Dr. W. Junk, Der Haag, 1–21.

Troll, C. 1971. Landscape ecology (geoecology) and biogeocoenology—a terminological study. Trans. by E.M. Yates. *Geoforum* **8**:43–46.

Turchin, P. 1993. Chaos and stability in rodent population dynamics: evidence from a non-linear time-series analysis. *Oikos* **68**:167–172.

Turner, M.G. 1989. Landscape ecology: the effect of pattern on process. *Annu. Rev. Ecol. Syst.* **20**:171–97.

Turner, M.G., G.J. Arthaud, R.T. Engstrom, S.J. Hejl, J. Liu, S. Loeb, and K. McKelvey. 1995. Usefulness of spatially explicit population models in land management. *Ecol. Appl.* **5**:12–16.

Ulanowicz, R.E. 1997. *Ecology, the Ascendent Perspective.* Columbia University Press, New York.

Ulanowicz, R.E. and L.G. Abarca-Arenas. 1997. An informational synthesis of ecosystem structure and function. *Ecol. Model.* **95**:1–10.

Ulanowicz, R.E. and J. Norden. 1990. Symmetrical overhead in flow networks. *Int. J. Syst. Sci.* **21**(2):429–437.

Urban, D.L., R.V. O'Neill, and H.H. Shugart. 1987. Landscape ecology. *BioScience* **37**(2):119–127.

Väisänen, R.A., O. Järvinen, and P. Rauhala. 1986. How are extensive, human-caused habitat alterations expressed on the scale of local bird populations in boreal forests? *Ornis Scand.* **17**:282–292.

Van den Aarsen, L.F.M. 1994. Randvoorwaarden voor natuurlijke kwatiteit in pleistocene zand--gebieden, een onderzoek vanuit de per-sistentietheorie in het perspectief van planning. Thesis, Wageningse Ruimte-lijke Studies 10.

van der Maarel, E. 1982. Biogeographical and landscape-ecological planning and nature reserves, in *Perspectives in Landscape Ecology.* S.P. Tjallingii and A.A. de Veer, Eds. Centre for Agricultural Publishing and Documentation, Wageningen, Netherlands, 227–235.

Vander Zouen, W.J. and D.K. Warnke. 1995. Deer population goals and harvest management environmental assessment. Wisconsin Department of Natural Resources. Madison.

Vaughan, T.A. 1986. *Mammalogy.* Saunders College Publishing, Philadelphia, PA.

VeldKamp, A. and L.O. Fresco. 1996. CLUE. A conceptual model to study the conversion of land use and its effects. *Ecolog. Modelling.* **85**:253–270.

Verboom, J., A. Schotman, P. Opdam, and J.A.J. Metz., 1991. European nuthatch metapopulations in a fragmented agricultural landscape. *Oikos* **61**:149–156.

Verschuur, G.L. 1996. *Impact!* Oxford University Press, Oxford, UK.

Virkkala, R. 1990. Ecology of Siberian tit Parus cinctus in relation to habitat quality: effects of forest management. *Ornis. Scand.* **21**:139–146.

Virkkala, R. 1991. Spatial and temporal variation in bird communities and populations in north-boreal coniferous forests. *Oikos* **62**:59–66.

Wahlberg, N., A. Moilanen, and I. Hanski. 1996. Predicting the occurrence of endangered species in fragmented landscapes. *Science* **273**:1536–1538.

Wallace, J.B., S.L. Eggert, J.L. Meyer, and J.R. Webster. 1997. Multiple trophic levels of a forest stream linked to terrestrial litter inputs. *Science* **277**:102–104.

Walters, M.J. 1992. *A Shadow and a Song.* Chelsea Green Publishing, Post Mills, VT.

Walters, C.J. and C.S. Holling. 1990. Large-scale management experiments and learning by doing. *Ecology* **71**:2060–2068.

Wardle, D.A., O. Zackrisson, G. Hörnberg, and C. Gallet. 1997. The influence of island area on ecosystem properties. *Science* **277**:1296–1299.

Webb, P.-N. and D.M. Harwood. 1991. Terrestrial flora of the Sirius Formation: its significance for late Cenozoic glacial history. Review 1987, *Antarct. J. U.S.* **12**:84–88.

Webster, J.R., J.B. Wallace, and E.F. Benfield. 1995. In river and stream ecosystems. C.E. Cushing. K.W. Cummins, and G.W. Minshall, Eds. Elsevier Science Ltd., Amsterdam, Netherlands.

Wegner, J.F. and G. Merriam. 1979. Movements by birds and mammals between a wood and adjoining farmland habitat. *J. Appl. Ecol.* **16**:349–357.

Wennergren, U., M. Ruckelshaus, and P. Karieva. 1995. The promise and limitations of spatial models in conservation biology. *Oikos* **74**:349–356.

Werner, H.W. 1975. The biology of the Cape Sable sparrow. Report to U.S. Fish and Wildlife Service, Frank M. Chapman Memorial Fund, The International Council for Bird Preservation, and U.S. National Park Service, Homestead, Florida. 215 pp.

Whitcomb, R.F., C.S. Robbins, J.F. Lynch, B.L. Whitcomb, M.K. Klimkiewicz, and D. Bystrak. 1981. Effects on forest fragmentation on avifauna of the eastern deciduous forest, in *Forest Island Dynamics in Man-Dominated Landscapes*. L. Burgess and D.M. Sharpe, Eds. Springer-Verlag, New York, 125–205.

Whittaker, R.H. and S.A. Levin. 1977. The role of mosaic phenomena in natural communities. *Theor. Popul. Biol.* **12**:117–139.

Wiens, J.A. 1985. Vertebrate responses to environmental patchiness in arid and semi-arid ecosystems, in *Natural Disturbance: an Evolutionary Perspective*. S.T.A. Pickett and P.S. White, Eds. Academic Press, New York, 169–193.

Wiens, J.A. 1992. What is landscape ecology, really? *Landscape Ecol.* **7**(3):149–150.

Wiens, J.A. 1995. Landscape mosaics and ecological theory, in *Mosaic Landscapes and Ecological Processes*. L. Hansson, L. Fahrig, and G. Merriam, Eds. Chapman & Hall, London, 1–26.

Wiens, J.A., N.C. Stenseth, B. Van Horne, and R.A. Ims. 1993. Ecological mechanisms and landscape ecology. *Oikos* **66**:369–380.

Wijnhoven, A.L.J. 1982. Welcome, in *Perspectives in Landscape Ecology*. S.P. Tjallingii and A.A. de Veer, Eds. Centre for Agricultural Publishing and Documentation, Wageningen, Netherlands, 5–6.

Wilcove, D.S. 1985. Nest predation in forest tracts and the decline of migratory songbirds. *Ecology* **66**:1211–1214.

Wilcox, B.A. and D.D. Murphy. 1985. Conservation strategy: the effects of fragmentation on extinction. *Am. Nat.* **125**:879–887.

Williams, G.C. 1966. *Adaptation and Natural Selection: a Critique of Some Current Evolutionary Thought*. Princteon University Press, Princeton, NJ.

Willson, M.F., S.M. Gende, and B.H. Marston. 1998. Fishes and the forest. *BioScience* **48**(6):455–462.

Wilson, E.O. 1975. *Sociobiology: the New Synthesis*. The Belknap Press. Cambridge, MA.

Wilson, D.S. 1997. Biological communities as functionally organized units. *Ecology* **78**(7):2018–2024.

Wilson, D.S. 1992. Complex interactions in metacommunities, with implications for biodiversity and higher levels of selection. *Ecology* **73**:1982–2000.

Winsor, C.P. 1934. Mathematical analysis of growth of mixed populations. *Cold Spring Harbor Symp. Quant. Biol.* **2**:181–189.

Wolff, W.F. 1994. An individual-oriented model of a wading bird nesting colony. *Ecol. Model.* **72**:75–114.

Woodroffe, R. and J.R. Ginsberg. 1998. Edge effects and the extinction of populations inside protected areas. *Science* **280**:2126–2128.

Wright, G.M. and B.H. Thompson. 1935. Fauna of the national parks of the USA: Wildlife management in the national parks. Fauna Series 2, U.S.A. Government Printing Office, Washington, D.C.

Yano, E. 1978. A simulation model of searching behavior of a parasite. *Res. Popul. Ecol.* **20**:105–122.

Zlotin, R.I. and K.S. Khodashova. 1980. *The Role of Animals in Biological Cycling of Forest-Steppe Ecosystems.* Dowden, Hutchinson, & Ross, Stroudsburg, PA.

Zonneveld, I.S. 1988. Landscape ecology and its application, in *Landscape Ecology and Management.* M.R. Moss, Ed. Polyscience Publications, Montreal, 3–15.

Zonneveld, I. 1994. Landscape ecology and ecological networks, in *Landscape Planning and Ecological Networks.* E.A. Cook and H.N. van Lier, Eds. Elsevier, New York, 13-26.

Index

F

M

N